Lecture Notes in Mechanical Engineering

Lecture Notes in Mechanical Engineering (LNME) publishes the latest developments in Mechanical Engineering—quickly, informally and with high quality. Original research or contributions reported in proceedings and post-proceedings represents the core of LNME. Volumes published in LNME embrace all aspects, subfields and new challenges of mechanical engineering.

To submit a proposal or request further information, please contact the Springer Editor of your location:

Europe, USA, Africa: Leontina Di Cecco at Leontina.dicecco@springer.com
China: Ella Zhang at ella.zhang@cn.springernature.com
India, Rest of Asia, Australia, New Zealand: Swati Meherishi at swati.meherishi@springer.com

Topics in the series include:

- Engineering Design
- Machinery and Machine Elements
- Mechanical Structures and Stress Analysis
- Automotive Engineering
- Engine Technology
- Aerospace Technology and Astronautics
- Nanotechnology and Microengineering
- Control, Robotics, Mechatronics
- MEMS
- Theoretical and Applied Mechanics
- Dynamical Systems, Control
- Fluid Mechanics
- Engineering Thermodynamics, Heat and Mass Transfer
- Manufacturing Engineering and Smart Manufacturing
- Precision Engineering, Instrumentation, Measurement
- Materials Engineering
- Tribology and Surface Technology

Indexed by SCOPUS, EI Compendex, and INSPEC.

All books published in the series are evaluated by Web of Science for the Conference Proceedings Citation Index (CPCI).

To submit a proposal for a monograph, please check our Springer Tracts in Mechanical Engineering at https://link.springer.com/bookseries/11693.

Agnimitra Biswas · Dipankar Bhanja ·
Om Prakash Singh · Shankar Chakraborty
Editors

Recent Advancements in Artificial Intelligence and Optimization

Select Proceedings of ICRAME 2025

Springer

Editors
Agnimitra Biswas
Department of Mechanical Engineering
National Institute of Technology Silchar
Silchar, Assam, India

Dipankar Bhanja
Department of Mechanical Engineering
National Institute of Technology Silchar
Silchar, Assam, India

Om Prakash Singh
Department of Mechanical Engineering
Indian Institute of Technology Varanasi
Varanasi, Uttar Pradesh, India

Shankar Chakraborty
Department of Production Engineering
Jadavpur University
Kolkata, West Bengal, India

ISSN 2195-4356 ISSN 2195-4364 (electronic)
Lecture Notes in Mechanical Engineering
ISBN 978-981-95-9267-8 ISBN 978-981-95-9268-5 (eBook)
https://doi.org/10.1007/978-981-95-9268-5

This Springer imprint is published by the registered company Springer Nature Singapore Pte Ltd.
The registered company address is: 152 Beach Road, #21-01/04 Gateway East, Singapore 189721, Singapore

Contents

Assessment of Grid-Integrated Electric Vehicle Charging Station Based on Floating Solar PV: A Review

Mukul Banerjee, Sourav Saha, and Sukanta Roy

Abstract This study aims to demonstrate the significant potential and cooperation between electric vehicles (EVs) and solar energy. Examining the viability of a national energy system that runs entirely on solar power and electric cars shows this possibility. Although photovoltaic (PV) solar energy is already a significant source of energy on a worldwide scale, energy storage is necessary to balance periods of high and low production because of its unpredictability. Large-scale PV plants, however, are unsustainable due to their high area requirements, especially in nations with higher population densities. A tempting way to harness solar energy while preserving the priceless land is to install photovoltaic plants on bodies of water, such as lakes, dams, reservoirs, rivers, canals, and ponds. Floating solar photovoltaics (FSPV) is another name for this kind of plant. Installing floating photovoltaic (FSPV) systems is a good substitute for solar energy consumption when land access is restricted. Concurrently, there is a global movement in the transportation sector to switch from internal combustion engines to electric vehicles. Since cars are actually immobile 95% of the time, the EV battery can use vehicle-to-grid (V2G) technology to control the intermittent FPV source upon connecting the vehicle to the grid. This renewable EVCS also reduces energy expenses by 74%. Some of the highlights of the review are the environmental impacts, potential future advancements, cost-effectiveness, FPV deterioration, and the ambient temperature of solar panels on water bodies to boost and improve the effectiveness of EVCS.

Keywords Renewable energy · Floating solar photovoltaic (FSPV) systems · Electric vehicles charging

---

M. Banerjee (✉) · S. Saha · S. Roy
Department of Mechanical Engineering, National Institute of Technology Mizoram, Aizawl, India
e-mail: banerjeeindia.10@gmail.com

S. Roy
Department of Mechanical Engineering, Curtin University Malaysia, Miri, Sarawak, Malaysia

A. Biswas et al. (eds.), *Recent Advancements in Artificial Intelligence and Optimization*,
Lecture Notes in Mechanical Engineering,
https://doi.org/10.1007/978-981-95-9268-5_1

1 Introduction

Although various fuels are used as the primary source of energy generation globally, coal remains the largest contribution at 38%. Focus has shifted toward renewable resources, especially in the last 10 years, due to conventional fuels' finite supply, rising costs, and negative environmental effects. Consequently, with a 14.5% rise in net energy output, the proportion of renewable resources ranked third, surpassing both natural gas (25%) and coal (31%). One of the various renewable energy sources is the solar photovoltaic systems. This innovative and conceptual study aims to accomplish two goals. First, to demonstrate the significant potential and complementary nature of solar energy and electric cars (EVs). Second, to illustrate this possibility by researching the viability of establishing a statewide charging network that is exclusively dependent on EVs and FPV. The renewable energy source with the most promise worldwide is unquestionably solar energy, with a potential of more than a thousand times larger than global energy consumption [1]. PV is the energy source with the greatest growth rate in the world, with power installations increasing by 30% annually between 2012 and 2019 [2]. With such a rapid growth rate, PV will overtake hydropower in less than five years and surpass wind energy in 2020 to become the world's largest installed renewable energy power capacity [3]. The enormous potential and importance of FSPVs are demonstrated by the fact that Japan has the largest installation of floating PV plants worldwide, accounting for over 50% of all existing floating PV plants. Between 2016 and 2018, the number of FSPVs installed rose sharply from 132 to 1097 [4]. FPV operation is mentioned in the aforementioned Fig. 1. This review's specific objective is to analyze and assess the feasibility and environmental and economic factors of installing floating solar photovoltaic plants connected to the EVCS, considering the FSPV electrical power potential, payback period, and CO_2 emission reduction.

2 An Analysis Comparing FPV and LPV

Photovoltaic systems installed on the ground are often very effective solar installations. The solar modules are in a rack next to the equipment field assistant's use and the ground. Ballast accessory holders, such as steel-reinforced concrete bases, which employ vital components to secure solar PV modules in place, do not require grounding. The solar cell system is lightweight and simple to assemble, and this development method performs well in hard-to-excavate locations like confined soils [5]. Figure 2 mentions a comparison of the Daily Data for [A] (Floating Solar Panels) and [B] (Rooftop Solar Panels). The module's ideal temperature means power generation efficiency is poor on hot summer days on land. On the other hand, the water's temperature reduces the surrounding air temperature, perhaps increasing its effectiveness.

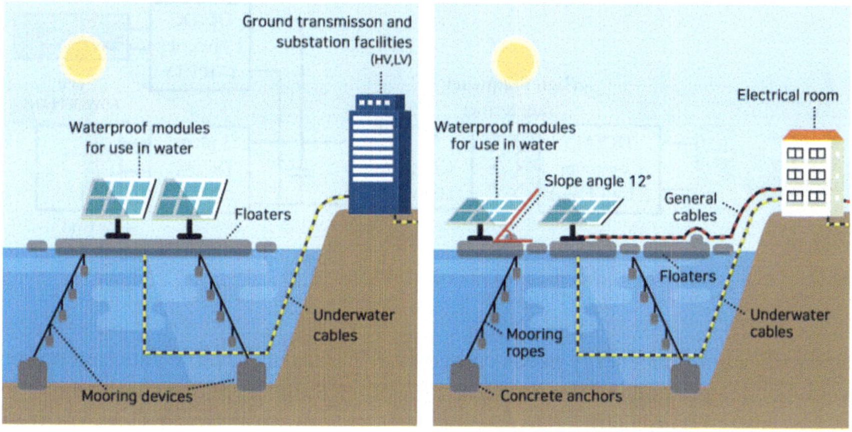

[1] [2]

Fig. 1 FPV working process

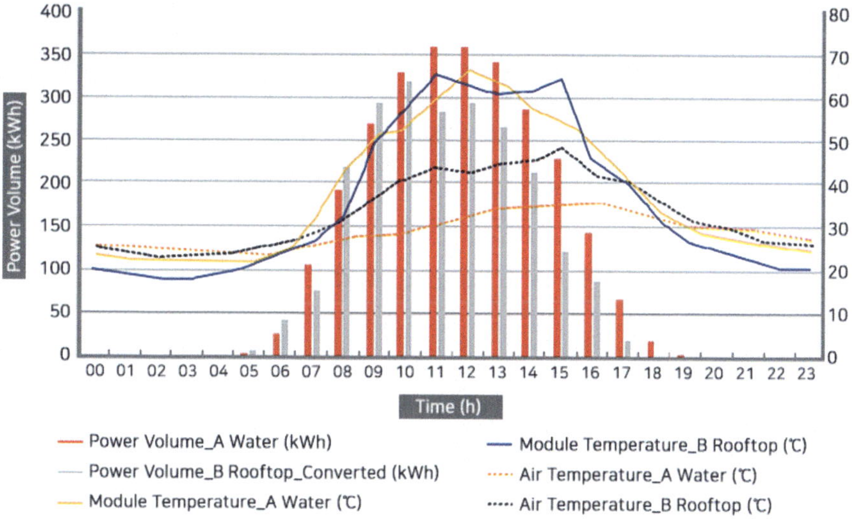

Fig. 2 Comparison of daily data for A (Floating solar panels) and B (Rooftop solar panels) [6]

3 Observation of EV Charging Incorporated into FPV

The predicted dramatic increase in EV use will unavoidably overburden the electrical system. Adding renewable energy sources to the grid is one way to address the problem. Solar energy is one possible source that could be considered in this FPV. Much research looks at the long-term, extensive impacts of FPV solar energy to

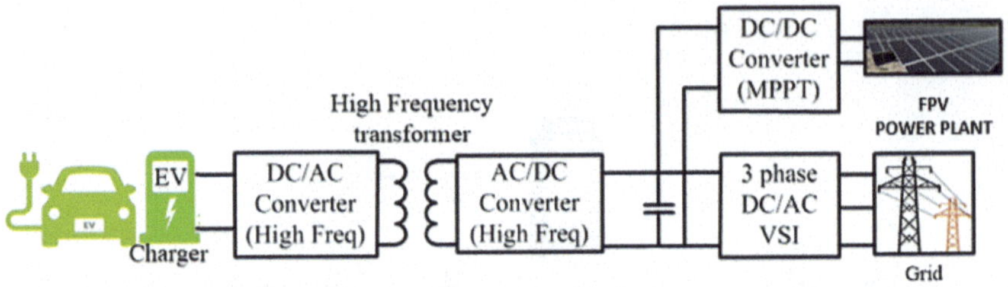

Fig. 3 EV charging power generation integrated to FPV

satisfy the additional energy needs caused by EV charging [7]. EV charging power generation integrated to FPV is shown in Fig. 3. Both the energy generated by the FPV stations and the energy utilized to power the EV stations are observed. However, they are noticeably out of balance because solar FPV energy is intermittent and unavailable at night. Energy storage devices or the grid may be used as a backup energy source to correct this imbalance [8]. A number of aspects of system component control and protection are explained. More specifically, the system's overall functioning depends on two essential parts of the battery and charger management systems (CMS and BMS, respectively), which are in charge [9]. The system design demonstrates that the charger module comprises numerous reversible DC-DC converter units. The expectation that the EV charger will someday include V2G functionality is why the bidirectional topology is used. A high switching frequency is appropriate, given that the converter operates between 75 and 250 kHz [8].

4 Connected to FPV-EV Charging and Energy Storage Systems

Rapid charging stations can less impact the MV grid's distribution when energy storage solutions are employed. These gadgets can lower the maximum amount of electricity and improve the network. Additionally, the voltage level can be raised via ESS. If the voltage reduction along the lines is too great, this feature requires voltage control [10]. By storing the power produced to operate massive FPV plants, an energy storage (ESS) system can reduce the voltage and frequency instability and eliminate peaks [11]. With a 60,000 KW/h capacity, one of the biggest battery storage facilities in the world was constructed in Hokkaido, Japan, to provide direct FPV to the grid. A 32-MWh Li-Ion BESS is being built in Southern California to control frequency and voltage close to a 4,500,000 KW wind farm [12]. The one-system energy storage structure uses a single ESS and locally produced renewable energy resources to provide electricity. Lithium batteries are the most widely used and effective storage technology since these systems only require one type of storage, are usually less costly, and are simpler to build [13]. Figure 4 shows a hybrid micro-grid charger model for battery electric vehicles (BEVs).

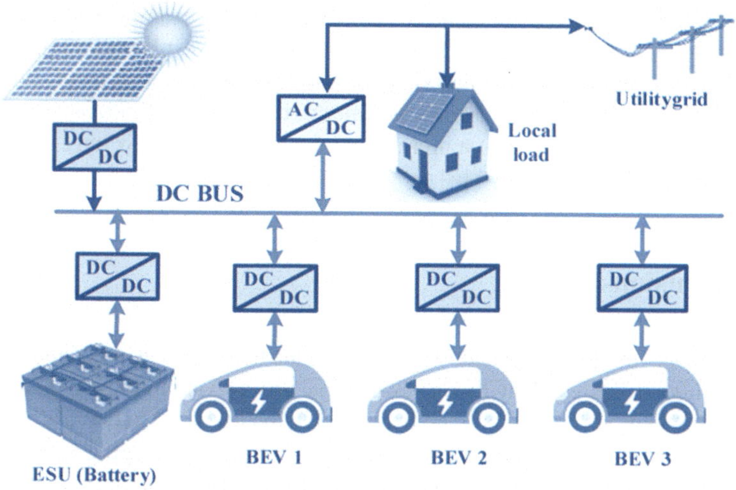

Fig. 4 Hybrid micro-grid battery electric vehicle (BEV) charging station model

5 The Importance of EVCSs Based on FPV

- FPV-based EVCSs can be connected to energy storage systems to improve resource efficiency and reduce emissions.
- The development of a DC micro-grid driven by FPV that can effectively balance electricity to satisfy load demand and charge electric cars.
- Micro-grid regulation conditions are optimized for various settings using fuzzy logic control and Sparrow search algorithms (SSA).
- A fuzzy SSA composite controller can be used to manage DC bus power, even when load uncertainty and irradiance volatility exist.
- This paper proposes a sustainable and reasonably priced energy infrastructure. The cost reductions that come from using controlled charging for electric cars (EVs) are covered in [13], also mentioned in Table 1.

Table 1 Competitiveness data on costs of the complete concept of EV fast charging: the distinction between LPV, FPV, and non-renewable electricity [15–17]

	Non-renewable energy	LPV	FPV
Location	Mahindra -Eva Mall Charging Station. Bengaluru, Karnataka,	Neel-Nirjan Dam of Bakreswar, West Bengal,	Harapanahalli, Karnataka
Capacity	50 KW	2000 KW	10,000 KW
Electricity rates (INR/KW/h)	8.49	5.04	2.19
Initial cost (INR)	16,49,000–19,39,000	2,70,00,000	94,00,000

- Nine of India's most prominent coastal locations are examined in this study. The states selected for the EVCS implementation are West Bengal, Tamil Nadu, Andhra Pradesh, Kerala, Karnataka, Maharashtra, Goa, Gujarat, and Odisha.
- AI's capacity to evaluate enormous volumes of data is revolutionizing transportation and opening the door to a more sustainable future; EVCSs have effectively incorporated AI technology [14].

6 Conclusions

This review covers the integration of floating solar PV energy sources into the EV charging infrastructure. Many literary works that are relevant to the region are thoroughly evaluated. Several criteria are published in the literature to classify and evaluate the research on FPV and EV integration, including EV charging technologies, different charging levels, and electric vehicle charging station standards, guidelines, different charging procedures, and certain AI applications. The power demand, inadequate infrastructure for charging, the usage of intelligent energy management systems, the robustness and adaptability of power structures, and charging via portable technologies are all topics of discussion. FPV integration with EVCSs is essential for power generation systems in the future worldwide market.

7 Future Scope

To maximize the advantages of solar PV on electric vehicles, the top goals must be to develop energy storage systems and improve management strategies. By using solar electricity on water bodies, these systems might lessen conflicts over land use, enhance panel performance, and provide a dependable and effective energy source for EV chargers. This connection encourages the growth of alternative energy sources while reducing the overall carbon footprint of EV charging.

References

1. Perez R, Perez M (2015) Update 2015-a fundamental look at the energy reserves of the planet. SHCP Solar Update, p 62
2. Fraunhofer Institute for Solar Energy System (ISE) (2020) Photovoltaics report. Freiburg, Germany. https://www.ise.fraunhofer.de/content/dam/ise/de/documents/publications/studies/Photovoltaics-Report.pdf
3. IHA (International Hydropower Association) (2020) Hydropower status report-sector trends and insights
4. Floating Solar Market Report, World Bank Group (2020). https://documents.worldbank.org/curated/en/579941540407455831/pdf/Floating-Solar-Market-Report-Executive-Summary.pdf

5. Yousuf H, Khokhar MQ, Zahid MA, Kim J, Kim Y, Cho EC, Yi J (2020) A review on floating photovoltaic technology (FPVT). Curr Photovolt Res 8(3):67–78
6. Wong wang S &T Co. Ltd. http://www.wksnt.com/eng/bbs/content.php?co_id=pro07_en
7. Al-Ghussain L, Alrbai M, Al-Dahidi S, Lu Z, Lee P (2024) Techno-economic and environmental assessment of solar-based electrical vehicles charging stations integrated with hydrogen production. J Clean Prod 434:140219
8. Suvvala J, Dhananjayulu C, Kotb H, Elrashidi A (2024) Integration of renewable energy sources using multiport converters for ultra-fast charging stations for electric vehicles: an overview. Heliyon
9. Sheldon TL, Dua R (2019) Measuring the cost-effectiveness of electric vehicle subsidies. Energy Econ 84:104545
10. Sangeetha E, Ramachandran V (2022) Different topologies of electrical machines, storage systems, and power electronic converters and their control for battery electric vehicles—a technical review. Energies 15(23):8959
11. Elsisi M, Amer M, Su CL (2023) A comprehensive review of machine learning and IoT solutions for demand-side energy management, conservation, and resilient operation. Energy 128256
12. Ma CT (2019) System planning of grid-connected electric vehicle charging stations and key technologies: a review. Energies 12(21):4201
13. Weis A, Jaramillo P, Michalek J (2014) Estimating the potential of controlled plug-in hybrid electric vehicle charging to reduce operational and capacity expansion costs for electric power systems with high wind penetration. Appl Energy 115:190–204
14. Aldhanhani T, Abraham A, Hamidouche W, Shaaban M (2024) Future trends in smart green iov: vehicle-to-everything in the era of electric vehicles. IEEE Open J Veh Technol
15. Yendaluru RS, Karthikeyan G, Jaishankar A, Babu S (2020) Techno-economic feasibility analysis of integrating grid-tied solar PV plant in a wind farm at Harapanahalli, India. Environ Prog Sustain Energy 39(3):e13374
16. Goswami A, Sadhu P, Goswami U, Sadhu PK (2019) Floating solar power plant for sustainable development: a techno-economic analysis. Environ Prog Sustain Energy 38(6):e13268
17. Mahindra XUV400 Variants. https://auto.mahindra.com/suv/xuv400/X400.html

Detection and Classification of Cotton Leaf Diseases: A Comprehensive Approach

Bhumika Prajapati and Vipul Dabhi

Abstract In India, Leaf diseases are a main cause of reduced crop productivity. In this paper, we focus on the detection and classification of cotton leaf diseases. We captured the infected leaf diseased images from the agricultural field Anand. We considered totally 105 images of three diseases namely Alternaria leaf spots (35 images), Bacterial blight (35 images), and Nitrogen deficiency (35 images). The captured images were used as image pre-processing. We considered the different background of the images such as cream, sky blue, with shadow and without shadow etc. So, the background of the images was removed from the images as required for the image pre-processing step. Modified TSAI technique was used for background removal. After images were pre-processed, image segmentation step was performed. In image segmentation, color image segmentation and binary segmentation techniques were performed. Using image color segmentation, green pixels were masked and omitted from the image. Otsu method was used for binary image segmentation. A binary image was used to mask the image, and the mask image was used to image and color texture features. The features were extracted and classified using a library support vector machine—LIBSVM multiclass SVM toolbox. With a cost value of 12 and gamma value of 0.03 the technique provided 96% training and 93% testing accuracy.

Keywords Image Pre-processing · Image segmentation · Leaf diseases · Otsu thresholding · Support vector machine (SVM)

B. Prajapati
Information Technology Department, Madhuben and Bhanubhai Patel Institute of Technology, The Charutar Vidhya Mandal University (CVM), New VVNagar, India
e-mail: bsprajapati@mbit.edu.in

V. Dabhi (✉)
Information Technology Department, Dharamsinh Desai University (DDIT), Nadiad, India
e-mail: Vipul.k.dabhi@gmail.com

A. Biswas et al. (eds.), *Recent Advancements in Artificial Intelligence and Optimization*, Lecture Notes in Mechanical Engineering,
https://doi.org/10.1007/978-981-95-9268-5_2

1 Introduction

Some of the common causes of plant diseases are caused by fungal and viral infections. A virus-inflicted disease allows viruses to enter the plant via an injury which either stunts the plant's growth or is a simple plant or leaf injury. An example of a viral disease is the Cotton leaf curl virus. Some bacterial diseases include infection by bacteria which can attack stems and leaves. Bacterial blight is one example of a bacterial infection. A plant can be infected by all kinds of fungi and that type of infection is called a fungal infection. It is treated with fungicides. A fungal infection that results in a grey discoloration of the leaves is termed as grey mildew [1].

Disease identification from the leaf is one of the crucial tasks in disease detection. The main problem in disease detection is each disease has different characteristics from each other such as differences in color, shape, and size of disease spot. Alternaria leaf spots look as tiny, round spots that are brown, grey-brown, or tan in color. These spots have varying sizes between 1-and 10 mm. Bacterial blight differs from this because it has a water-soaked area that enlarges to form reddish angular spots which are approximately 1 mm in diameter [2, 3, 15].

There are different image processing concepts available to detect diseases such as image pre-processing, image segmentation, feature extraction [2, 4, 5]. The extracted features are further fed into classifiers to detect diseases. As for the image preprocessing phase, one performs background extraction, and video resizing and applies histogram equalization. It is then possible to proceed to image segmentation analysis. During this process, methods like canny and sobel, k-means clustering, Otsu thresholding, as well as other techniques are utilized. From the processed image, the features of the object are color, shape, and texture. These features are extracted and classified with other relevant data by a classifier like support vector machines or a neural network for disease detection purposes. There are some previously proposed works for disease detection but these works have limitations of low accuracy results [5] and a smaller set of images in diagnosing the disease [6].

This paper has eight sections. Section 2 covers various kinds of cotton leaf diseases. Major types of cotton leaf diseases are also discussed in the same section. Furthermore, this chapter presents a discussion on different leaf diseases which relate to our work. The fourth part explains the results obtained from reviewed studies. This is how I intend to do my research work as explained in the V section above. This section provides an outline for how we plan on carrying out this study. In conclusion, future work is outlined in this section as well (Sect. 7).

2 Types and Characteristics of Cotton Leaf Diseases

There are a total of 14 types of diseases that can affect cotton plants. Diseases that mainly occurred on the cotton leaf are Alternaria leaf spot, Bacterial Blight, and Nitrogen deficiency. These illnesses are described together with their symptoms, stages of incidence, and occurrence.

2.1 *Alternaria Leaf Spot*

- Occurrence: When humidity is very high, this disease is found most frequently.
- Symptoms: Brown, round or irregular spots on leaf with crack and without cracked centers.
- Affecting Factors: High humidity, Intermittent rains, Reasonable temperature.
- Disease occurrence stage: Last stage (Fig. 1).

2.2 *Bacterial Blight*

- Occurrence: It reduces the photosynthetic activity of the plant.
- Symptoms: The disease appears first as small, round, water-soaked, raised spots and the lesions gradually became irregular in shape, turn brown to deep black in color and occur in the centerAffecting Factors: Bacterial infection.
- Disease occurrence stage: Middle to last stage (Fig. 2).

Fig. 1 Alternaria leaf spot

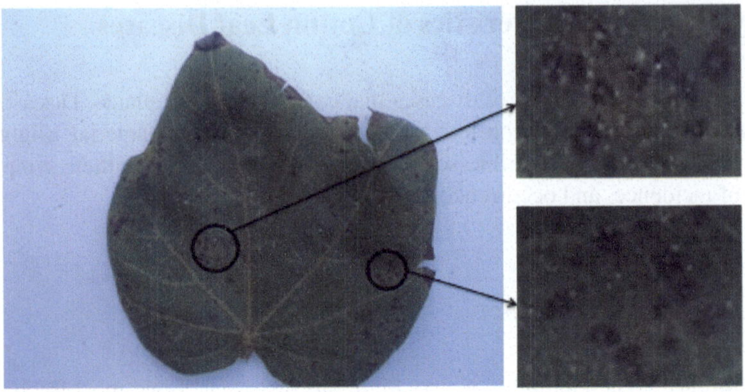

Fig. 2 Bacterial blight

2.3 Nitrogen Deficiency

- Occurrence: A shortage of nitrogen can reduce yields.
- Symptoms: Yellowish green leaf with irregular red spots.
- Affecting Factors: If soil does not contain adequate amount of nitrogen.
- Disease occurrence stage: Last stage (Fig. 3).

3 Findings of Literature Review

Through the survey of detection of cotton leaf diseases in our paper [7], we were unable to locate any common databases on cotton leaf disease. So, we collected photos of cotton leaves from the field near Anand and we created our own. Many writers have used digital cameras and cell phones to take photos of diseased images

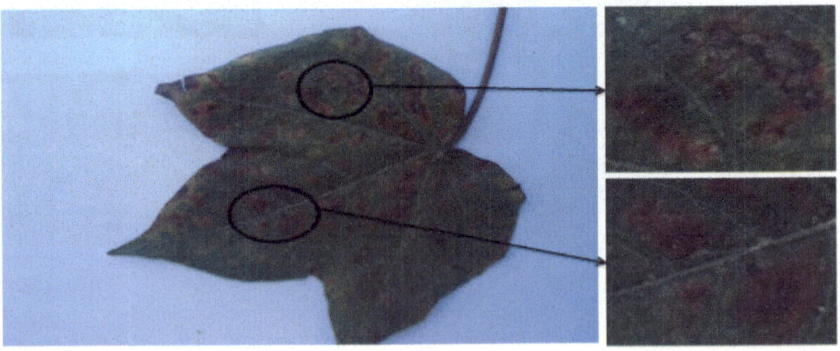

Fig. 3 Nitrogen deficiency

[5]. Using photos taken with mobile devices has the drawback that it requires additional pre-processing operations like scaling and picture restoration. As a result, we used a digital camera to take pictures of cotton leaves, which produces higher quality photographs.

Image Pre-processing is the first step in detecting the diseases afflicted on the cotton leaves [8]. Gaussian noise and salt and pepper noise are usually present in images and practitioners have used median filter to get rid of such noises from the leaf image. As stated by authors, a Histogram equalization has also been used as a technique to improve the sharpness and in imagery's Image preprocessing [9].

In the report the background is removed off the already processed image [7]. Background elimination can be reached in various techniques like contrast enhancement, changing color space, and image segmentation [10]. A contrast enhancement adds pixels to an image through the full range of tones in the image so as to make sections that are too dark and too light clearer. Another technique to aid in the removal of backgrounds is thresholding when the image is converted to another color space like (HSV, Cmyk). The H in HSV refers to the pure color of the image and hence achieving accurate results when processing that component H is challenging. While using CMYK some color components can be troublesome, so results are often inaccurate. To eliminate the background, segmentation technique will be used to separate the image into foreground and background pixels.

The approach to eliminate segments of the background was through camera subject segmentation which was carried out along with Otsu thresholding application. Instead of using the segmentation approach, on deleting Otsu segmentation masking technique would be more appropriate since it enables one to directly remove the background without segmentation [11].

In their work, Otsu and Nakashima developed an automated threshold selection method and one of its most popular applications is segmentation of images using its most utilized segmentation feature of Otsu thresholding. The purpose of Otsu's thresholding is to find a value that minimizes the average variance of the intensities in the background regions. Otsu thresholding is largely relevant in the diagnosis of some spotted diseases of cotton leaves including Bacterial blight, Alternaria leaf spot, and Leaf blight. On the contrary, k-means clustering is another widely used technique which segments an image based on the specified number of clusters [12]. The most important problem when using k-means is that the pixels are divided into two groups, one of which is for similar pixels and the other one is reserved for dissimilar ones. It has also been utilized in the detection of different types of diseases such as Myrothecium, Bacterial blight [13], and Alternaria [14], and even Cercospora leaf spot [1]. The drawback of using k-means technique is the randomly generated output. K-means technique is more than often curtsied for the results it generates, this is snake segmentation which is rarely used.

A framework for removing an object's outline from a noisy picture is called the active contour model. It is a method for minimizing energy use. Therefore, the Otsu segmentation technique produces good results when used to remove the sick area of the image [15]. The methodology section included a discussion of these techniques' experiments.

One of the main responsibilities in the disease detection process is feature extraction. The mean and standard deviation of the specified color space work well for color features when determining the average color value of the affected area. The primary traits that are helpful in detecting illnesses are those of shape and texture. Disease-related body components vary in size and shape [16]. The diameter, eccentricity, aspect ratio, area, and other shape characteristics can be utilized to determine the size of the affected area. The sick part's area is located using the area function. Aspect ratio is measured in terms of eccentricity. It is the major to minor axis ratio. To identify the texture of a diseased part, texture parameters like energy, entropy, dissimilarity, cluster shadow, and cluster prominence are crucial [17]. Features that add to misclassification and lengthen the classifier's computation time and do not convey meaningful information should be removed from the feature space. The feature reduction technique can be used to accomplish this.

For classification, support vector machines were adopted by numerous practitioners [3, 18–20]. Support vector machines are used with hyperplanes to categorize datasets. Neural networks have not been widely employed by authors to categorize diseases. The neural network technique requires a lot of time, and it can be challenging to determine which learning algorithm—such as feed forward or back-propagation neural networks—will produce the best results. SVM provides accurate results for classification in terms of accuracy and disease detection.

4 Our Approach to Detect and Classify Cotton Leaf Diseases

Figure 4 shows architecture of the proposed work. The steps are as follows: Image Acquisition, Image Pre-processing, Image Segmentation, Feature Extraction, and Classification.

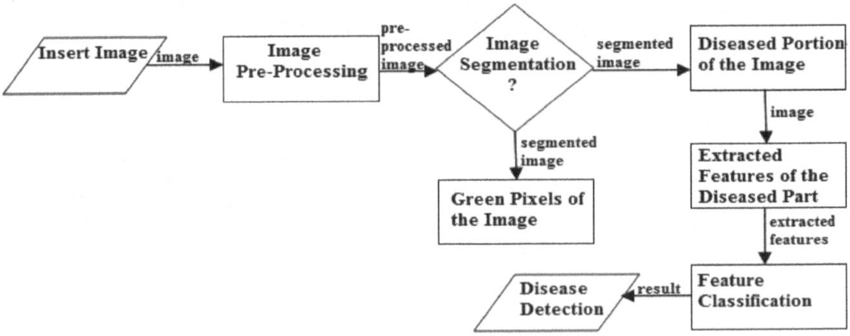

Fig. 4 Proposed work to detect and classify cotton leaf disease

4.1 Image Acquisition

Images were captured from Anand agricultural university, Anand. There were a total 180 diseased leaf images and 15 healthy leaf images. We have images of totally four types of diseases such as Alternaria, Bacterial blight, and Nitrogen deficiency. We have totally 80 images of Alternaria leaf images, 40 images of Bacterial blight, and 40 images of Nitrogen deficiency.

4.2 Image Pre-processing

Images were pre-processed by removing the background of images. Background of the images was removed using modified TSAI technique as shown in Fig. 11.

4.3 Image Segmentation

For image segmentation two techniques were used. For color image segmentation green pixels were masked and removed from the image as shown on Fig. 14. For binary image segmentation Otsu thresholding is applied on the color segmented image. Otsu segmentation converts RGB image into binary image using a threshold value. Inbuilt graythresh function was used for Otsu thresholding.

4.4 Feature Extraction

Color and binary picture segmentation yields features including color, texture, and shape. The mean and standard deviation values of the red, green, and blue components were retrieved for the purpose of extracting color features. Using various angles, the texture attributes contrast, autocorrelation, energy, entropy, and homogeneity were retrieved. In addition, cluster prominence and shade were identified as textural elements. Shape features were recovered from the binary segmented image, including area, diseased portion, number of disease spots, major and minor axis modes, roundness, eccentricity, elongation, diameter, minimum diseased area, and mode of the diseased region. 69 features in all were taken out of the segmented image that was color and binary whereby there were two color features, twelve form features, and seven texture features. Contrast, correlation, energy, and homogeneity were found with different offsets of the image, which totally account for 48 features (16 for each).

4.5 Classification

Features were classified using multiclass support vector machine. For multiclass svm LIBSVM tool was used. We considered three diseases for classification, namely Alternaria, Bacterial blight, and Nitrogen deficiency. We have total 105 images as training (35 for each disease) and 15 images as testing (5 for each).

5 Methodology of Proposed Work

In this section, the various steps for methodology of our proposed work are described in detail.

5.1 Image Pre-processing

We have tested our images totally on four background removal techniques.

5.1.1 Static Thresholding

First, we have tested static thresholding for background removal. Block diagram of static thresholding technique presented in Fig. 8. Background removal using static threshold, gives good results for background removal but in some cases, the shadow was not perfectly removed from the image. We must change the threshold value for some of the images.

5.1.2 Fuzzy C-means Thresholding

Fuzzy c-means thresholding was also used for background removal. Block diagram of fuzzy c-means technique presented in Fig. 9.

$$\text{Level} = \frac{(\max(data(label == 1)) + \min(data(label == 1)))}{2}$$

Fuzzy c-means technique gives better result in compared to static threshold. Fuzzy c-means removes the background and shadow of the image but in some cases when shadow is also present on the leaf portion it is also removed from the image which contains green part of the leaf image (Fig. 5).

Fig. 5 Results of fuzzy C-means thresholding

5.1.3 Background Removal Technique

In the presented technique taken from [33], the block diagram presented in Fig. 10. The presented technique easily removes the background and shadow from the image but while removing it, most of the green pixels images were also removed from the object while some of them contains diseased portion of the leaf. The presented technique only works for white background, it does not work for sky blue background of the image.

5.1.4 Modified TSAI Technique

Modified TSAI technique was also used for background removal. As shown in Fig. 11. ratio map was found based on the value of hue and intensity.

$$\text{Ratio map} = \frac{(Hue + 1)}{(I + 0.1)}$$

This technique works for both background white as well as sky blue to remove the background and shadow without losing green pixels of the image (Fig. 6).

Fig. 6 Results of technique presented in [7]

5.2 *Evaluation of Implemented Background Removal Techniques*

Evaluation of background removal techniques was carried out by comparing their results with ground truth of original images. We calculated TP, TN, FN and FP parameters by comparing Ground Truth image and background removed image. We also calculated accuracy and F-score measures (Fig. 7).

The relationship between GT and the background-removed image is known as true positive (TP). The combination of the background-removed picture and GT is the true negative (TN). False positive (FP) refers to the segmented area that GT does not overlap. The portion of GT that is missed is false negative (FN).

Following the findings of the FP, TP, TN, and FN below were values for true positive, true negative, false positive, and false negative. We concluded that the modified TSAI produced the greatest results in terms of F-score and accuracy based on the results of the TP, TN, FP, FN value, and F-score. The evaluation results of background removal strategies are shown in Tables 1, 2, and 3 (Figs. 8, 9, 10 and 11).

5.3 *Color Image Segmentation*

In this section, experimental results of various color image segmentation methods are shown. For image segmentation, two techniques were tested: K-means segmentation

Fig. 7 Results of modified TSAI technique

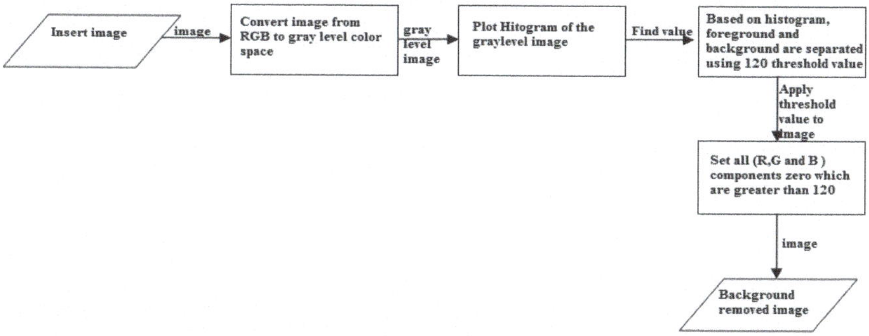

Fig. 8 Block diagram of static thresholding technique

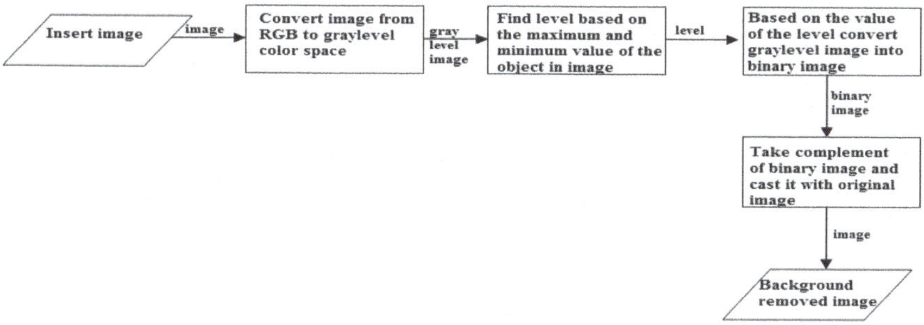

Fig. 9 Block diagram of fuzzy c-means thresholding

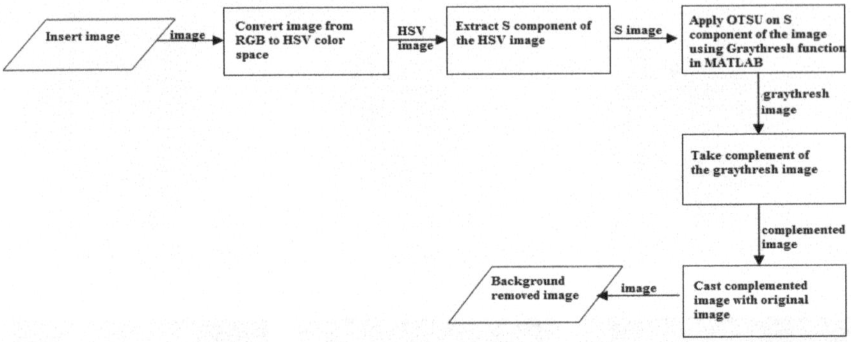

Fig. 10 Block diagram of background removal technique presented in [7]

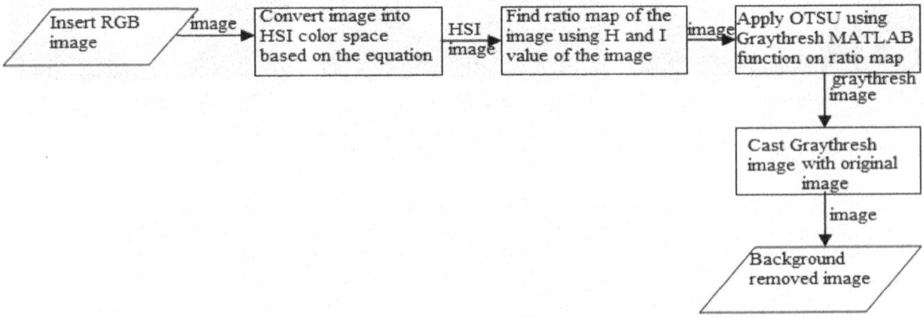

Fig. 11 Block diagram of modified TSAI technique

Table 1 Evaluations of background removal techniques with Alternaria_25 image

Alternaria_25	TP	TN	FN	FP	Accuracy	Precision	F-score
Static threshold	0.3642	0.3520	0.1951	0.0887	0.9730	0.9706	0.9637
Fuzzy c-means	0.3671	0.6074	0.0128	0.0125	0.9744	0.9668	0.9659
Technique taken from [7]	0.3595	0.6102	0.0203	0.0098	0.9693	0.9732	0.9582
Modified TSAI technique	0.3670	0.6082	0.0129	0.0117	0.9751	0.9689	0.9668

using CIELAB color space segmentation, and masking and removing of green pixels from the image using the threshold value.

Table 2 Evaluations of background removal techniques with Bacterial_23 image

Bacterial_23	TP	TN	FN	FP	Accuracy	Precision	F-score
Static threshold	0.2401	0.7412	0.0131	0.0054	0.9813	0.9480	0.9624
Fuzzy c-means	0.2400	0.7413	0.0130	0.0055	0.9813	0.9483	0.9624
Technique taken from [7]	0.2335	0.7452	0.0092	0.0120	0.9786	0.9620	0.9554
Modified TSAI technique	0.2404	0.7420	0.0123	0.0050	0.9825	0.9511	0.9648

Table 3 Evaluations of background removal techniques with Bacterial_29 image

Bacterial_29	TP	TN	FN	FP	Accuracy	Precision	F-score
Static threshold	0.4455	0.5031	0.0120	0.0393	0.9471	0.9736	0.9422
Fuzzy c-means	0.4609	0.4996	0.0155	0.0239	0.9599	0.9672	0.9577
Technique taken from [33]	0.4266	0.5051	0.0100	0.0581	0.9286	0.9769	0.9189
Modified TSAI technique	0.4718	0.5007	0.0144	0.0129	0.9723	0.9702	0.9713

5.3.1 K-means Segmentation Using CIELAB Color Space; Segmentation

Figure 12 presents the block diagram of K-means segmentation using CIELAB color space segmentation. The result is shown in Fig. 13. The result of k-means varies each time when we run the program, and the result does not give the exact disease portion needed for feature extraction.

5.3.2 Masking and Removing Green Pixels from the Image

A thresholding approach called "masking and removing green pixels" was used to mask and eliminate green pixels based on a predetermined threshold value. Apply a threshold value that is manually determined from our dataset and ranges from larger than 0.14 to less than 0.65. The green pixels in the image's Hue component are represented by these threshold values. Based on the outcomes of masking and deleting green pixels, we draw the conclusion that this method provides valuable

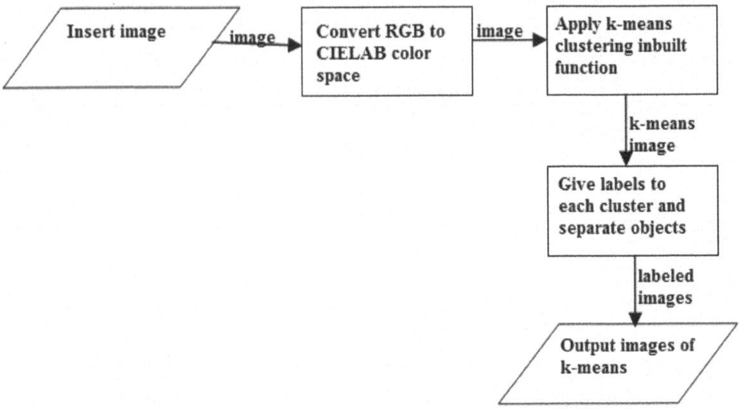

Fig. 12 K-means segmentation using CIELAB color space segmentation

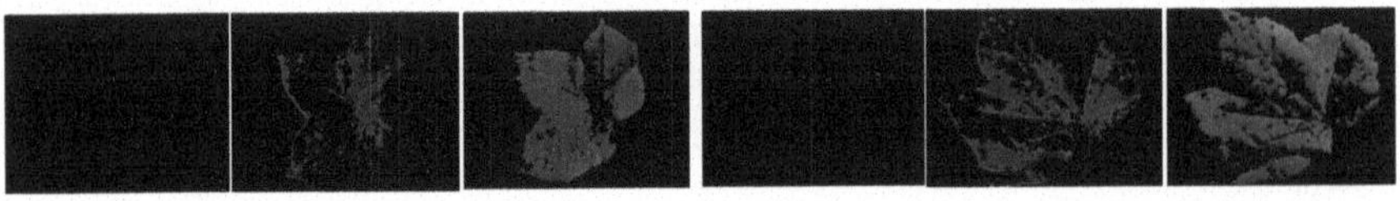

Fig. 13 Results of K-means segmentation using CIELAB color *space segmentation*

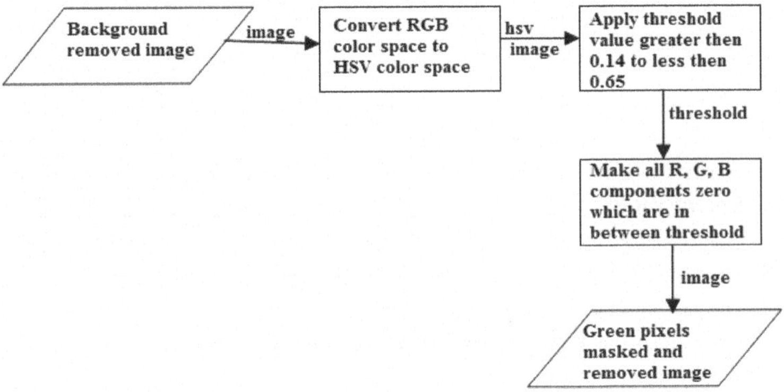

Fig. 14 Block diagram of masking and Removing green pixels from the image

information about the disease-related area of the image. Figure 14 displayed a block diagram showing how to mask and remove green pixels from the image (Fig. 15).

In order to identify the image's shape components, a binary segmented image was necessary. The color-segmented image was converted to binary using two binary routines. The color-segmented image was initially converted directly to binary. This binary conversion removes the sick portion of the image. Thus, we used the built-in Otsu function to convert the color segmented image into a binary image.

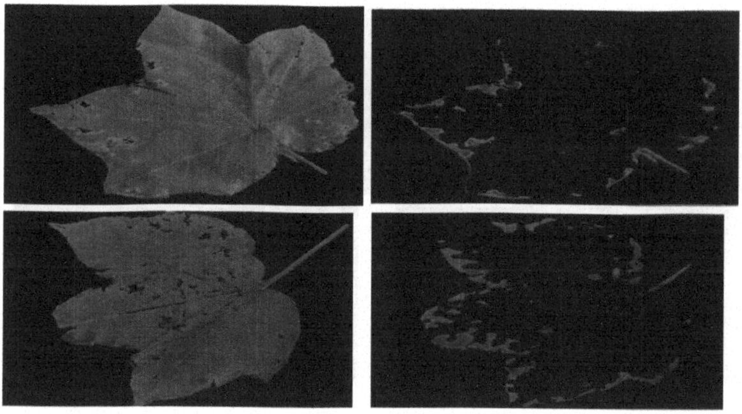

Fig. 15 Results of masking and removing green pixels from the image

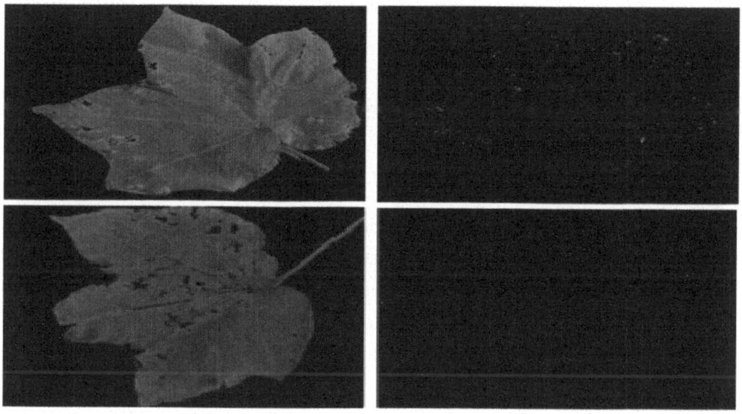

Fig. 16 Results of inbuilt binary function

5.3.3 Inbuilt Binary Function (im2bw)

The output of the built-in binary function im2bw is shown in Fig. 16. The built-in Matlab im2bw function for binary segmentation does not provide precise illness information needed to extract disease features. The precise diseased area of the picture is not shown.

5.3.4 Otsu Thresholding

When Otsu thresholding is performed with the built-in Matlab function graythresh, the outcome is displayed in Fig. 17. This function determines the level at which two sections were segmented, and it applies the grayscale function to the image based on

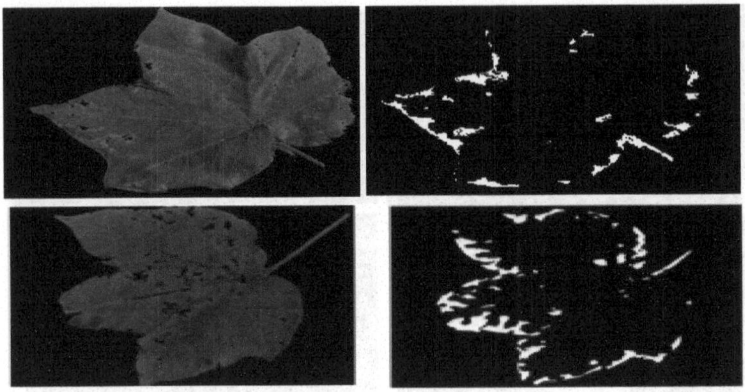

Fig. 17 Results of Otsu thresholding technique

that level. The precise diseased region was available in the color segmented image, thanks to Otsu thresholding.

5.4 Feature Extraction

The segmented and masked image was used to extract features like color, shape, and texture. The color segmented image was used to extract color features. The afflicted portion's mean and standard deviation values were retrieved for color characteristics. From the RGB picture, which consists of red, green, and blue components, the mean and standard deviation were extracted. The hue, saturation, and value components of the color segmented image was also extracted from the HSV image to extract color features. Table 4 displays color characteristics like the red and blue components' standard deviation values. Applying the mean function yielded the mean, and the std2 function yielded the standard deviation.

Aspect ratio, eccentricity, diameter, maximum and lowest diseased areas, mode of the infected region, major and minor axis modes, elongation, roundness, and area of the diseased portion are examples of shape attributes. Twelve features in all were taken out of the binary segmented picture. The retrieved form attributes, including illness portion, number of disease spots, aspect ratio, eccentricity, diameter, elongation, and roundness, which are displayed in Table 5. Shape characteristics were retrieved using the built-in Matlab function called regionprops.

$$\text{Disease portion} = \mathit{Area\ of\ one\ disease\ spot/Total\ area}$$

$$\text{Roundness} = \frac{4 * 3.14 * (\mathit{Total\ area})}{\mathit{Total\ perimeter} * \mathit{Total\ perimeter}}$$

Table 4 Extracted color features

Image name	Red standard deviation value	Blue standard deviation value
Alternaria_1	0.4661	0.7414
Alternaria_2	0.1104	0.3087
Alternaria_3	0.3564	0.4578
Bacterial_1	0.4336	0.0827
Bacterial_2	0.5836	0.5967
Bacterial_3	0.6404	0.5130
Nitrogen_1	0.3489	0.4329
Nitrogen_2	0.2642	0.4630
Nitrogen_3	0.3312	0.5656

Table 5 Extracted shape features

Image name	Disease portion	Aspect ratio	Eccentricity	Roundness
Alternaria_1	0.0160	0.2032	0.2128	0.1214
Alternaria_2	0.0032	0.0000	0.0175	0.0612
Alternaria_3	0.0119	0.1977	0.0953	0.1606
Bacterial_1	0.1271	0.2186	0.3948	0.1645
Bacterial_2	0.0112	1.0000	0.0000	0.3253
Bacterial_3	0.0505	0.1434	0.5140	0.0787
Nitrogen_1	0.1235	0.2781	0.2717	0.1905
Nitrogen_2	0.4324	0.1418	0.9034	0.0969

From the grayscale image, texture characteristics including contrast, autocorrelation, energy, entropy, homogeneity, kurtosis, skewness, and smoothness were retrieved. The color, shape, and texture features that were extracted are displayed in the tabular form below. The recovered texture properties, including contrast, energy, entropy, auto-correlation, cluster shadow, and cluster prominence, are displayed in Table 6. The color segmented image was first transformed from RGB to grayscale color space. Then, textures were recovered using GLCM (Gray level Co-occurrence matrix). From the grayscale image, texture characteristics including contrast, autocorrelation, energy, entropy, homogeneity, kurtosis, skewness, and smoothness were retrieved. Using the GLCM_feature1 function, autocorrelation, entropy, kurtosis, and skewness were extracted. Using the built-in GLCM function, features like contrast, correlation, energy, and homogeneity were retrieved.

After extracting all color, shape and texture features all features were normalized by finding the minimum and maximum value of the features. Min max normalization was applied on the features to make all features in one range [4].

Table 6 Extracted texture features

Image name	Contrast	Auto-correlation	Cluster shade	Cluster prominence	Image name	Contrast	Auto correlation	Cluster shade	Cluster prominence
Alternaria_1	0.0470	0.5841	0.0430	0.6096	Bacterial_2	0.0281	0.1650	0.0081	0.8007
Alternaria_2	0.0032	0.6147	0.0060	0.8363	Bacterial_3	0.0907	0.6631	0.0787	0.3750
Alternaria_3	0.0299	0.6178	0.0276	0.6510	Nitrogen_1	0.2215	0.7090	0.1964	0.2602
Bacterial_1	0.1132	0.6693	0.0683	0.2224	Nitrogen_2	0.3902	0.8199	0.2498	0.0484

$$\text{Min}-\text{max normalization} = \frac{\textit{Original feature value} - \min(\textit{all features row wise})}{\max(\textit{all features row wise}) - \min(\textit{all features row wise})}$$

5.5 *Classification*

The multiclass support vector machine was used for feature classification. The tool LIBSVM was used for multiclass svm. A total of 105 photos—35 for each disease—are used for training, and 15 images are used for testing (five for each). Tests of classification accuracy were conducted using varying numbers of characteristics. In Tables 7 and 8, we evaluated the accuracy of nine color features, eleven form features, and nine texture features.

In the second attempt in Table 8, we tested accuracy with 2 color features, 11 shape features, and 8 texture features.

In third attempt Table 9, the accuracy was tested with 2 color features, 11 shape features, and 56 texture features in which 48 feature values such as contrast, correlation, energy, and entropy were extracted with different offsets.

Table 7 Different C and gamma value results with 29 features

C	G	Fold	CV accuracy (%)	Training accuracy (%)	Testing accuracy (%)
1	1	5	78	92	87
1	0.5	10	77		
2	0.125	5	79	92.27	89
2	0.125	10	80		
128	0.125	5	79	93	89
128	0.125	10	80.95		

Table 8 Different C and gamma value results with 69 features

C	G	Fold	CV accuracy (%)	Training accuracy (%)	Testing accuracy (%)
1	0.5	5	87	87	80
1	0.5	10	89		
2	0.125	5	87	89	90
2	0.125	10	89		
128	0.3	5	90	96	93
128	0.3	10	89.95		

6 Proposed User Interface of Our System

We made a user interface which detects the disease name from the diseased image (Figs. 18 and 19).

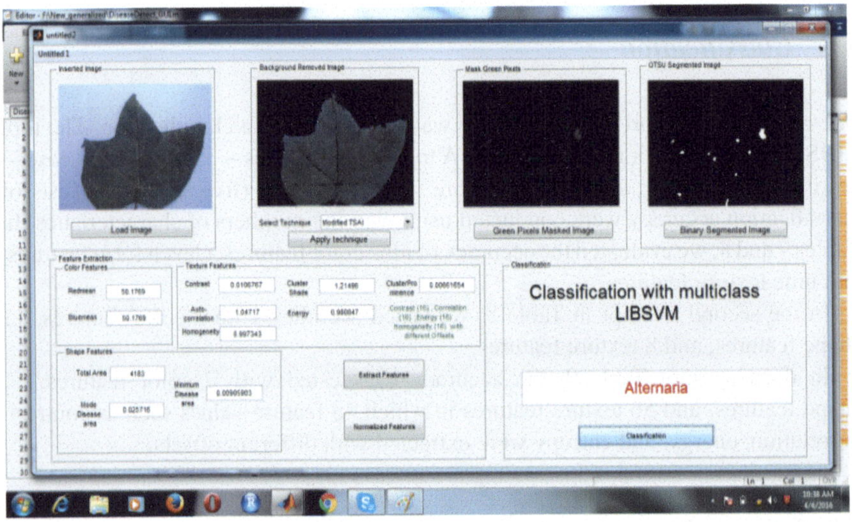

Fig. 18 User interface of alternaria disease

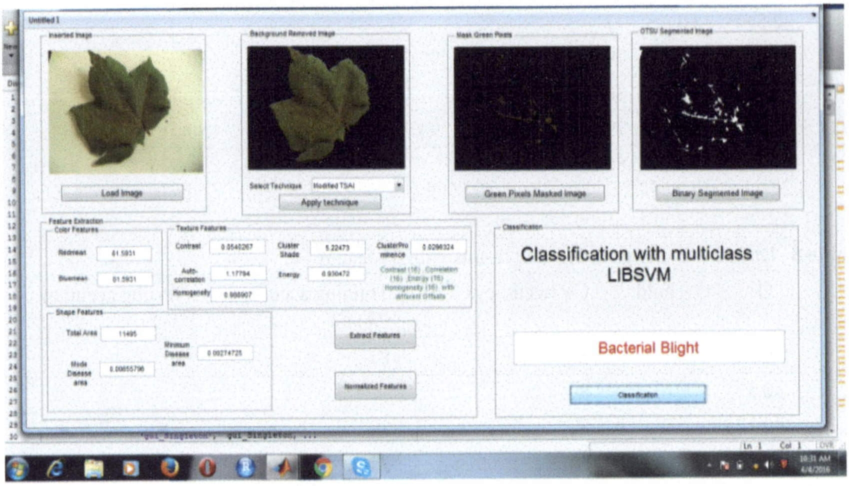

Fig. 19 User interface of bacterial disease

7 Result Comparison

Reference number and title	Techniques used for classification	Accuracy
3. B. Adsule and J. Bhattad	SVM and NN	80 to 90% approx
7. P. Warne and S. Ganorkar	K-means clustering	80%
8. P. Rothe and R Kshirsagar	SVM and NN	90%
11. P. Revathi and M. Hemalatha	ANN	85 to 91% approx
23. S. Arivazhagan, R. Newlin Shebiah et al	HSV color model with SVM	90%
Our proposed methodology	SVM with modified Otsu segmentation	93%

7.1 Conclusion

In this study, we have aimed to identify and classify the diseases that occur in cotton leaves. We developed a new approach that makes the detection and categorization of these diseases by examining the features of different cotton leaf diseases and by evaluating several disease detection and image segmentation techniques. Our methodology employed a modified TSAI background removal methodology followed by color and binary picture segmentation, where the dataset consisted of 105 images. With a radial basis kernel function, the color, texture, and shape data recovered were fed into the multiclass LIBSVM classifier, which yielded 96% training accuracy and 97% testing accuracy. Future studies could focus on developing new background removal techniques in uniform and non-uniform backgrounds as well as early-stage systems for this disease.

References

1. Types of Cotton diseases. http://louisianacrops.com/
2. Vishnu S, Ranjithram A (2015) Plant disease detection using leaf pattern: a review. Int J Innov Sci Eng Technol 2(6)
3. Revathi P, Hemalatha M (2012) Classification of cotton leaf spot diseases using image processing edge detection techniques. In: Emerging trends in science, engineering and technology (INCOSET), international conference. IEEE, pp 169–173
4. Lisin D, Mattar M, Blaschko M, Learned-Miller E, Benfield M (2005) Combining local and global image features for object class recognition. IEEE, Computer vision and pattern recognition-workshops
5. Revathi P (2012) Advance computing enrichment evaluation of cotton leaf spot disease detection using image edge detection. In: Computing communication & networking technologies (ICCCNT), third international conference. IEEE, pp 1–5

6. Adsule B, Bhattad J (2015) Leaves classification using SVM and NN for disease detection. Int J Innov Res Comput Commun Eng 3(6):5488–5495
7. Prajapati BS, Dabhi VK, Prajapati HB (2016) A survey on detection and classification of cotton leaf diseases. In: 2016 International conference on electrical, electronics, and optimization techniques (ICEEOT), Chennai, India, pp 2499–2506. https://doi.org/10.1109/ICEEOT.2016.7755143
8. Naik D, Vyavahare A (2015) Disease detection of cotton crop using image processing technique: a survey. Int J Res Appl Sci Eng Technol (IJRASET) 3(VI)
9. Warne P, Ganorkar S (2015) Detection of diseases on cotton leaves using k-mean clustering method. Int Res J Eng Technol (IRJET), pp 425–431
10. Revathi P, Hemalatha M (2012) Homogenous segmentation based edge detection techniques for proficient identification of the cotton leaf spot diseases. Int J Comput Appl (0975–888)
11. Arivazhagan S, Newlin Shebiah R, Ananthi S, Varthini S (2013) Detection of unhealthy region of plant leaves and classification of plant leaf diseases using texture features. Comm Int du Genie Rural (CIGR) J 15(1):211–217
12. Rathis A, Tanawal B, Shah V (2013) Image processing techniques for detection of leaf disease. Int J Adv Res Comput Sci Softw Eng 3(11)
13. Cercospora leaf spot of cotton. http://agritech.tnau.ac.in/crop_protection/cotton_diseases/cotton_d2.html
14. Bacterial Blight of cotton. http://agrihunt.com/articles/crop-diseases/angular-leaf-spot-of-cotton/
15. Gavhale K, Gawande U (2014) An overview of the research on plant leaves disease detection using image processing techniques. Int Organ Sci Res J Comput Eng (IOSR-JCE) 16(1):10–16
16. Shruti, Seth N (2014) Fungus/disease analysis in tomato crop using image processing techniques. Int J Comput Trends Technol (IJCTT) 13
17. Singh V, Varsha, Misra A (2015) Detection of unhealthy region of plant leaves using image processing and genetic algorithm. In: International conference on advances in computer engineering and applications (ICACEA)
18. Patil S, Zambre R (2014) Classification of cotton leaf spot disease using support vector machine. Int J Eng Res 3(4):1511–1514
19. Chaudhari V, Patil C (2014) Disease detection of cotton leaves using advanced image processing. Int J Adv Comput Res 4.2:653
20. Abhang D, Autade P, Bhokare S, Dawange P (2015) A survey on, analysis and detection of disease on cotton leaves. Int J Adv Res Innov Ideas Educ 4:236–243
21. Badnakhe M, Deshmukh P (2012) Infected leaf analysis and comparison by otsu threshold and k-means clustering. Int J Adv Res Comput Sci Softw Eng 2(3)
22. Rothe P, Kshirsagar R (2014) Svm-based classifier system for recognition of cotton leaf diseases. Int J Emerg Technol Comput Appl Sci (Ijetcas), pp 427–432
23. Rothe P, Kshirsagar R (2014) A study and implementation of active contour model for feature extraction : with diseased cotton leaf as example. Int J Curr Eng Technol 4(2):812–816
24. Characteristics of Cotton Diseases. http://www.daf.qld.gov.au/plants/field-crops-and-pastures/broadacre-field-crops/cotton/disease-management/
25. Cercospora Leaf Spot. http://www.mississippi-crops.com/
26. Kim J, Kim B, Savarese S (2012) Comparing image classification methods: K-nearest-neighbor and support-vector-machines. Ann Arbor 1001:48109–52122
27. Cotton Distress. http://cottonmarketnews.com/2015/05/27/india-record-rs-2-5k-cr-loss-seen-on-cotton-buying/
28. Vijayarani S, Vinupriya M (2013) Performance analysis of canny and sobel edge detection algorithms in image mining. Int J Innov Res Comput Commun Eng 1(8)
29. Dhaygude S, Kumbhar N (2013) Agricultural plant leaf disease detection using image processing. Int J Adv Res Electr Electron Instrum Eng 2(1):599–602
30. Kakade N, Ahire D (2015) A review of grape plant disease detection. Int Res J Eng Technol (IRJET) 02(05)

Optimization of Shell-and-Tube Heat Exchanger Performance Using CuO-Water Nanofluids: A Response Surface Methodology Approach

Mallesh Sanjeevannavar, Nagaraj Kelageri, Rajat Mangalwedhe, Kiran Haibatti, Suyash Desai, Chetan Takkekar, and Devratna Patil

Abstract This study employs the Response Surface Methodology (RSM) to optimize the performance of a shell-and-tube heat exchanger (STHE) utilizing CuO-water nanofluids. CuO nanoparticles, with an average diameter of 40 nm, were synthesized into nanofluids at volumetric concentrations ranging from 0.2 to 0.6% using a two-step method. The thermal and rheological properties were enhanced by optimizing the ultrasonication process to achieve a zeta potential of 46.5 mV, ensuring stability without sedimentation for over one month. The study demonstrated significant improvements in heat transfer performance compared to conventional configurations while maintaining operational feasibility regarding pressure drop. RSM-based optimization highlighted the influence of nonlinear interactions on performance metrics, such as effectiveness and fouling rates. These findings underscore the potential of CuO-water nanofluids in enhancing the efficiency of thermal management systems.

Keywords Shell-and-tube heat exchanger · CuO-water nanofluids · Response surface methodology · Heat transfer enhancement · Nanofluid stability · Zeta potential · Thermal management

---

M. Sanjeevannavar (✉) · N. Kelageri · R. Mangalwedhe · K. Haibatti · S. Desai · C. Takkekar · D. Patil
Department of Mechanical Engineering, KLE Technological University Belagavi Campus, Belagavi, Karnataka, India
e-mail: mallesh.sanjeevannavar@gmail.com

A. Biswas et al. (eds.), *Recent Advancements in Artificial Intelligence and Optimization*, Lecture Notes in Mechanical Engineering,
https://doi.org/10.1007/978-981-95-9268-5_3

1 Introduction

Heat exchangers are crucial in various industries, and with rising energy costs, improving their thermal efficiency is essential. The Shell and Tube Heat Exchanger (STHX) is widely used in sectors like petrochemical, chemical, and power generation due to its ability to handle high temperatures and pressures. However, optimizing its performance remains challenging due to flow arrangement, tube material conductivity, and fouling. To improve efficiency, strategies such as modifying tube geometry, using enhanced surfaces, and employing nanofluids are being explored for better performance. This study examines the heat transfer and pressure drop characteristics of conical spiral tube heat exchangers (STHEs) with Ag-HEG nanofluid and various turbulator designs in turbulent flow (Dean Numbers 2200–4200). Among three turbulator designs, the four-blade turbulator with 30 revolutions achieved the highest Nusselt number, enhancing heat transfer without excessive pressure drop [1]. A novel heat exchanger with continuous spiral baffles in corrugated tubes showed up to 55% improved friction and 40% better heat transfer, leading to 21% higher heat efficiency [2]. Optimization of STHEs with disc-doughnut baffles and dimpled tubes demonstrated significant improvements in heat transfer performance [3], while a slipcover-based performance model showed that baffle diameter and number affect heat transfer, enhancing performance by 32.35% and reducing mass flow rate by 29.18% [4]. In SHCTHEXs, ANN models reliably predicted heat transfer and pressure drops based on various parameters [5]. Passive enhancement methods for CO_2 exchangers identified twisted elliptical and eccentric helical tubes as optimal configurations for better performance and lower pressure drop [6]. Research on different tube cross-sections found that circular tubes achieved the highest Nusselt number, improving heat transfer by 62.5% [7]. Increasing inlet velocities and reducing tube spacing improved heat transfer in elastic tube bundles [8]. Studies on SiC and CNT nanofluids showed that CNTs outperformed SiC in heat transfer efficiency and pressure drop management [9]. Recent studies focus on improving heat exchanger efficiency using nanofluids and innovative designs, with experimental validation and optimization techniques offering valuable insights. Advanced methods like Response Surface Methodology (RSM) have enhanced performance optimization [10, 11]. However, a gap exists in combining RSM with CuO nanofluids, which could further improve efficiency. Addressing this gap promises advancements in energy efficiency and sustainability.

2 Material and Methods

2.1 Nanofluid Preparation and Its Properties

As shown in Fig. 1, CuO nanoparticles were used to prepare CuO-water nanofluids at concentrations ranging from 0.2 to 0.6%, kept below 1% to minimize costs and sedimentation (Fig. 2). The nanofluids were prepared using the two-step method, which requires techniques like surfactants, surface modification, and ultrasonic energy to prevent nanoparticle aggregation. An optimal ultrasonication time of 180 minutes yielded the best results, with a zeta potential of 46.5 mV and no sedimentation after one month. For industrial-scale production, each 10 L batch was prepared using a 32 kHz–160 W ultrasonic bath for 3 hours, without dispersants, to maintain fluid properties. The nanofluid properties are shown in Tables 1 and 2.

3 Experimentation

The experimental procedure involves setting up a shell-and-tube heat exchanger (STHE) shown in Fig. 3 and pictorial representation of shell tube tubes are shown in Figs. 4, 5, 6 and the baffle is shown in Fig. 7, and using CuO-water nanofluids for testing. The STHE is cleaned, and temperature gauges, flow meters, and pressure gauges are installed to monitor key parameters. Hot water is circulated on the shell side, while nanofluids flow on the tube side using high-pressure pumps. The water is preheated to the desired temperature, and flow rates are adjusted to achieve specific conditions, such as a mass flow rate of 0.4 kg/s and a Reynolds number of 4181.75. Inlet and outlet temperatures, flow rates, and pressure drops are recorded once the system stabilizes. The experiment is conducted for six nanofluid concentrations (0.2%, 0.372%, 0.368%, 0.46%, 0.54%, and 0.6%) and repeated for each

Fig. 1 CuO-nanoparticles

Fig. 2 CuO-nanofluid

Table 1 Properties of Cuo nanoparticle

S. no	Physical form	Powder
1	Purity	>99%
2	Average particle size	>100 nm
1	Bulk density	0.9 g/cm^3
2	SSA	60–80 m^2/g
3	CAS NO	1317-38-0
4	Morphology of nano particle	Near rod to spherical
5	Color	Dark Brown

Table 2 Thermophysical properties of Cuo nanofluids

Property	Cuo nanoparticle concentrations					
	0.20%	0.37%	0.37%	0.46%	0.54%	0.60%
Density (kg/m^3)	1002	1005	1005	1006	1007	1008
Viscosity (mPa·s)	0.92	1.05	1.04	1.08	1.12	1.15
Specific heat (J/kg·K)	4180	4135	4137	4120	4105	4090
Thermal conductivity (W/m·K)	0.62	0.69	0.68	0.71	0.74	0.76
Zeta potential (mV)	45	46.2	46	46.5	47	47.5
Prandtl number	6.8	6.5	6.5	6.4	6.3	6.2

concentration to ensure accuracy. Afterward, the data is analyzed to evaluate heat transfer rates, heat transfer coefficients, effectiveness, and pressure drops to assess STHE performance under varying nanofluid concentrations.

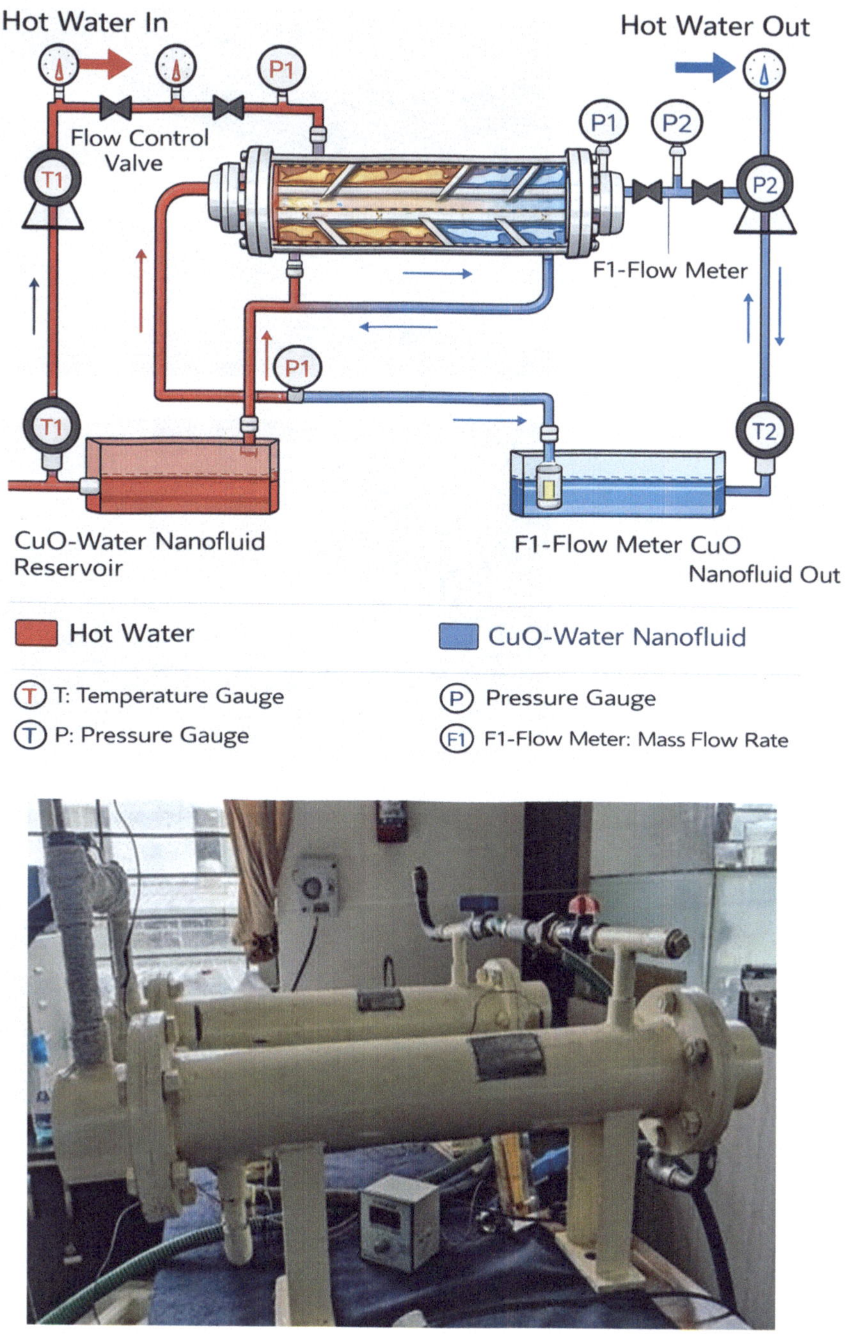

Fig. 3 Representation of shell and tube heat exchanger

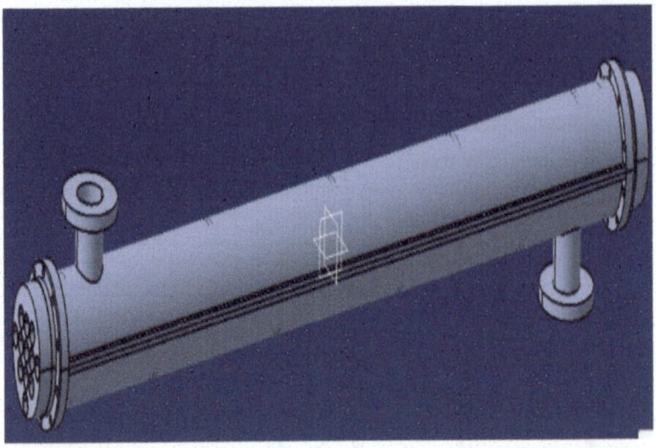

Fig. 4 Heat exchanger shell

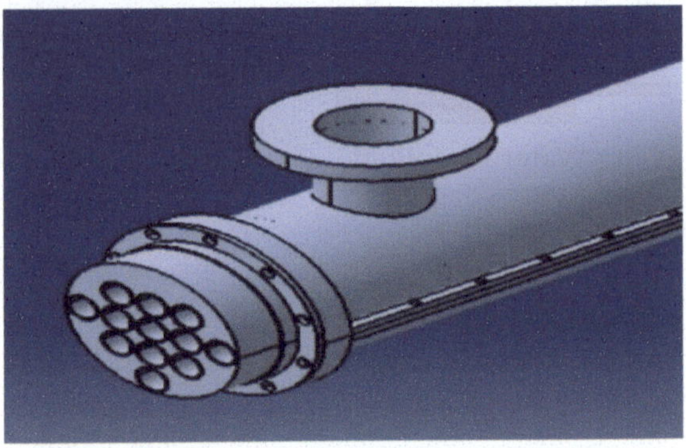

Fig. 5 Heat exchanger shell

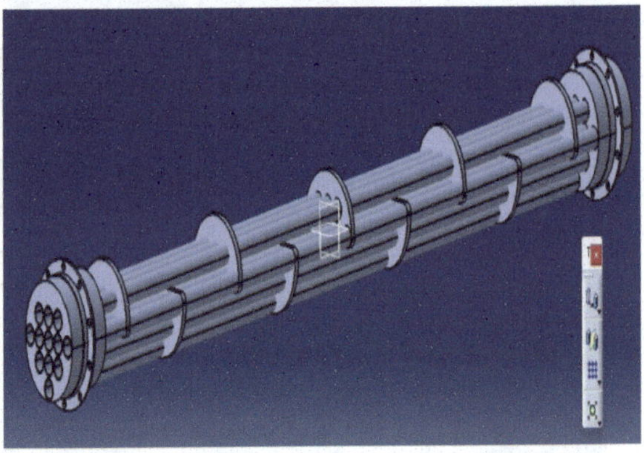

Fig. 6 Heat exchanger tube

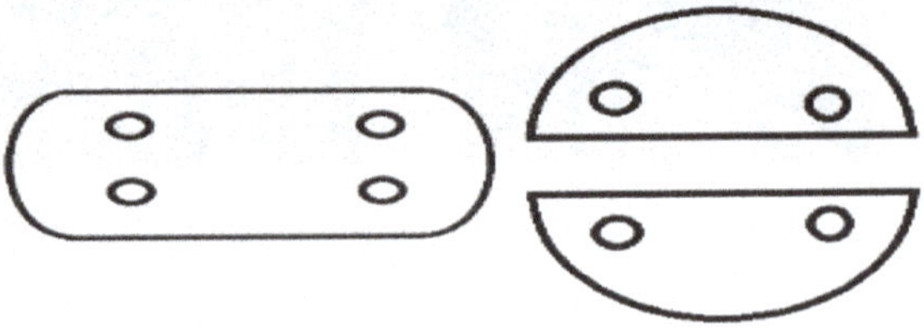

Fig. 7 Baffle plate

4 Response Surface Methodology

The current study extensively analyzed a shell-and-tube heat exchanger's performance using Response Surface Methodology (RSM), a robust statistical and mathematical tool for modelling and optimizing complex systems. The study provides significant insights into the factors influencing heat exchanger performance and their underlying relationships by systematically evaluating parameters such as specific heat, thermal conductivity, pressure drop, fouling rate, heat transfer rate, and effectiveness.

4.1 Specific Heat at Tube and Shell Side

The specific heat models shown in Figs. 10 and 11 for both the shell side and tube side were statistically significant ($p = 0.0415$), with strong contributions from the quadratic terms A^2 (connection type) and B^2 (mass flow rate), exhibiting p-values of 0.0114 and 0.0139, respectively, indicating a pronounced nonlinear relationship between the input variables (connection type and mass flow rate) and specific heat. The interaction term (AB) was not significant ($p = 0.9805$), suggesting that the individual effects of connection type and mass flow rate, along with their quadratic influences, dominate over any interaction effects. Both the predicted versus actual plot and 3D surface plots shown in Figs. 8 and 10 further confirm the accuracy and reliability of the model, making it a valuable tool for predicting specific heat under varying conditions. These findings provide key insights for optimizing heat exchanger performance by fine-tuning connection configurations and mass flow rates (Fig. 9).

4.2 Thermal Conductivity at Tube and Shell Side

The overall models shown in Figs. 13 and 15 for thermal conductivity on both the tube and shell sides were not statistically significant ($p > 0.05$), indicating limited predictive capability based on the primary terms and interactions. However, the quadratic terms A^2 and B^2 demonstrated significant influence, with p-values below 0.05 on both sides. These results as shown in Figs. 12 and 14 highlight that while the individual factors and their interactions have minimal impact, higher-order effects are critical in determining thermal conductivity. This underscores the importance of incorporating nonlinear terms in predictive models to capture the behavior of thermal conductivity under varying conditions accurately.

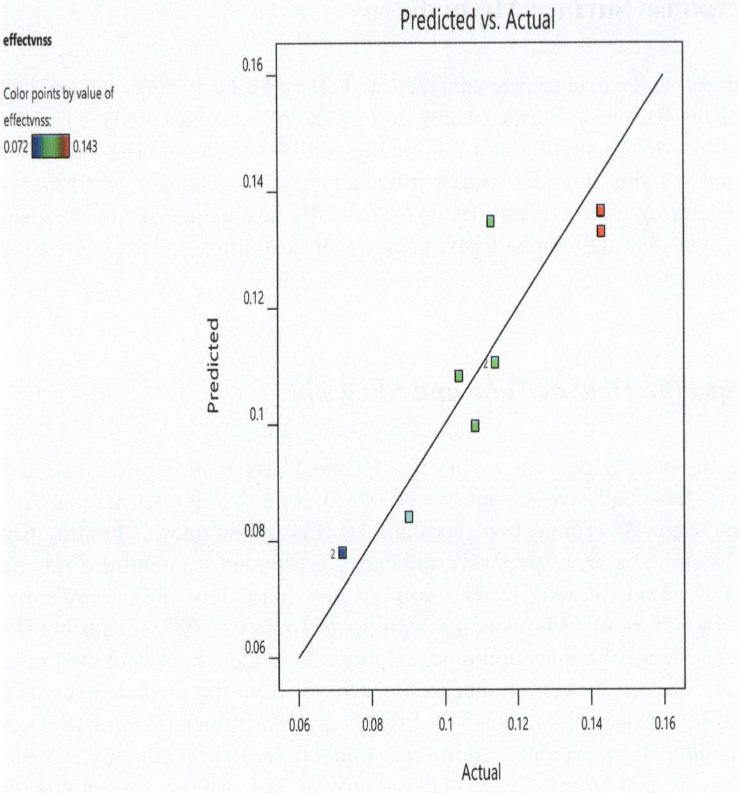

Fig. 8 Predicted versus actual specific heat at tube side

4.3 Pressure Drop at Tube and Shell Side

The pressure drop model shown in Fig. 17 for the tube side was highly significant (p = 0.0009), with mass flow rate emerging as the most influential factor (p = 0.0002), indicating its critical role in determining pressure drop. Additionally, the quadratic term A^2 significantly impacted (p = 0.0015), highlighting the importance of nonlinear effects shown in Fig. 15. In contrast, the shell-side model exhibited less overall significance as shown in Fig. 19, with notable contributions only from the quadratic terms A^2 and B^2, with p-values of 0.046 and 0.0345, respectively. These findings emphasize the dominant role of flow-related parameters in influencing pressure drop on the tube side. At the same time, nonlinear terms are crucial for capturing pressure drop behavior on the shell side shown in Fig. 16. This underscores the need for nonlinear terms in predictive models to represent pressure drop characteristics on both sides accurately (Fig. 18).

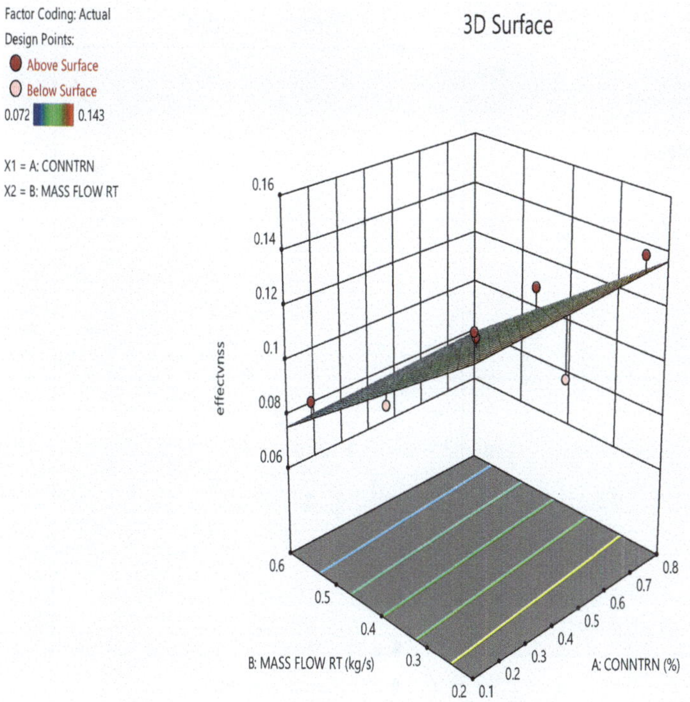

Fig. 9 Interactive effects between mass flow rates and concentration of Cuo Nanofluid on specific heat at tube side for RSM response surface (3D) plots

4.4 *Fouling Rate at Tube and Shell Side*

The fouling rate models shown in Figs. 21 and 23 for both the tube and shell sides were not statistically significant ($p > 0.05$), indicating that the current models lack comprehensive predictive power. However, the quadratic terms A^2 and B^2 were statistically significant ($p < 0.05$), highlighting their substantial influence on the fouling behavior as shown in Figs. 20 and 23. This suggests that fouling rates are predominantly governed by higher-order relationships rather than direct or interactive effects of connection type and mass flow rate. The non-significance of the overall model points to the need for incorporating additional variables, such as operating temperature and surface properties, to enhance the accuracy and comprehensiveness of fouling rate predictions. These findings emphasize the complexity of fouling behavior and the necessity of considering a broader range of factors in predictive modelling (Fig. 22).

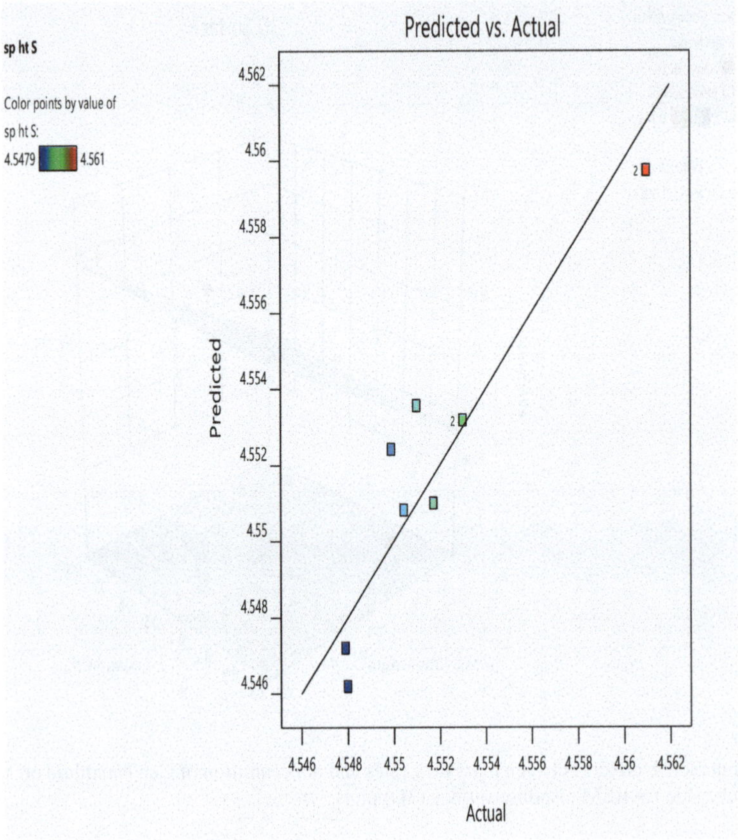

Fig. 10 Predicted versus actual specific heat at shell side

4.5 Heat Transfer Rate

The heat transfer rate model shown in Fig.25 was one of the most significant models in the study, with an overall p-value of 0.0011. Mass flow rate emerged as the dominant variable ($p = 0.0003$, $p = 0.0003$, $p = 0.0003$), demonstrating its critical role in enhancing heat transfer efficiency. Connection type had no significant influence ($p = 0.6971$, $p = 0.6971$, $p = 0.6971$), indicating that flow dynamics primarily govern heat transfer. The predicted versus actual plots shown in Figs. 24 and 3D response surfaces validate the robustness of the model, providing a reliable framework for optimizing heat exchanger performance.

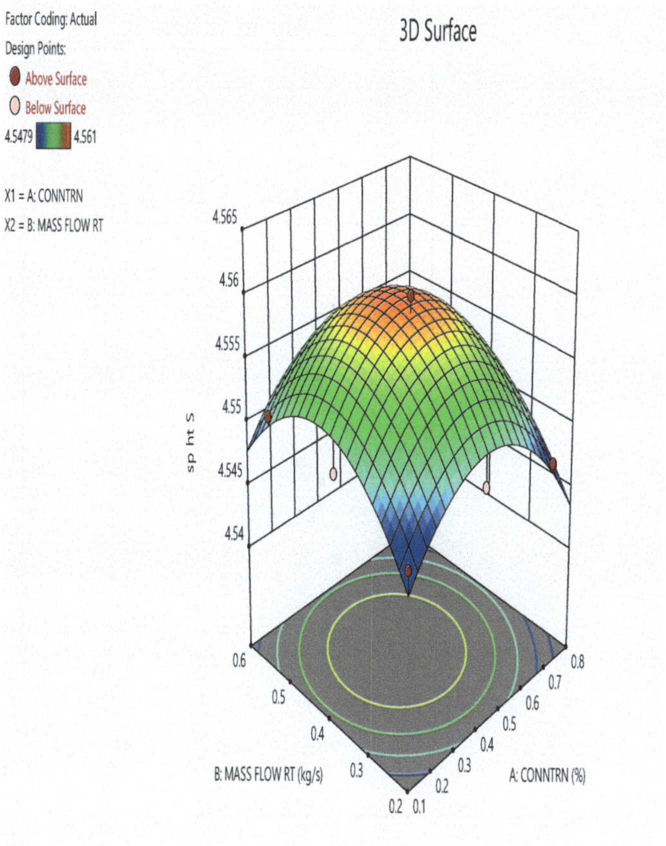

Fig. 11 Interactive effects between mass flow rates and concentration of Cuo Nanofluid on specific heat at shell side for RSM response surface (3D) plots

4.6 Effectiveness

The effectiveness model shown in Fig. 27 was also significant ($p = 0.0015$), with mass flow rate being the primary factor influencing system effectiveness ($p = 0.0005$, $p = 0.0005$, $p = 0.0005$). The connection type showed no notable effect ($p > 0.05$, $p > 0.05$, $p > 0.05$). These results align with the heat transfer rate analysis, reinforcing the idea that optimizing flow rates can significantly enhance the overall performance and effectiveness of the heat exchanger as shown in Fig. 26.

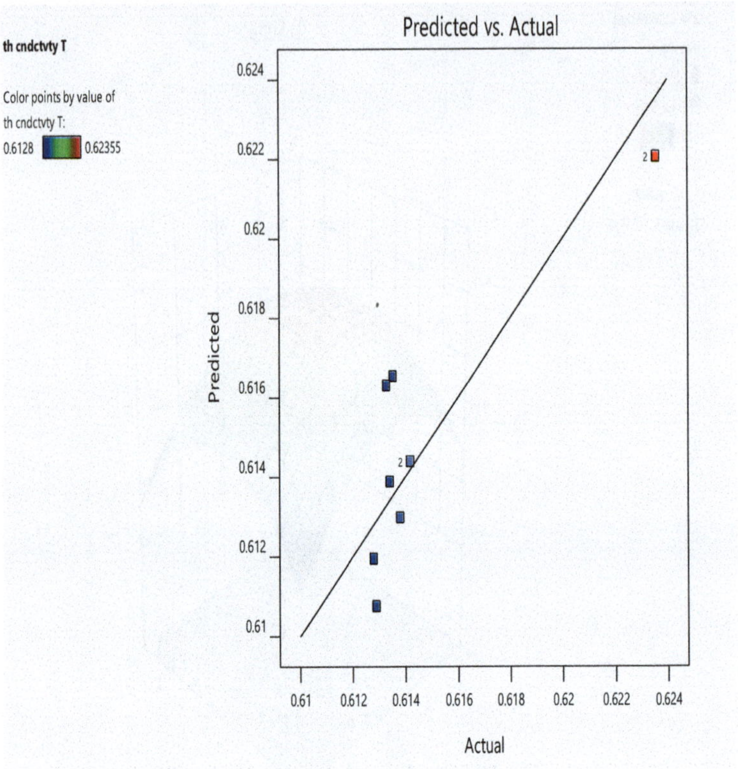

Fig. 12 Predicted versus actual thermal conductivity at tube side

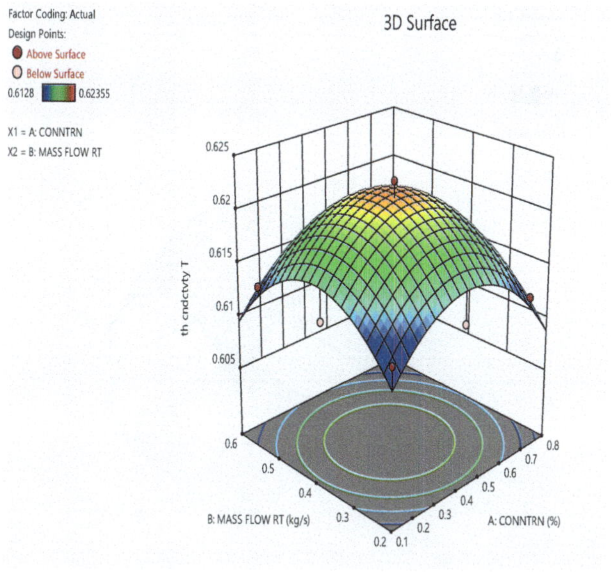

Fig. 13 Interactive effects between mass flow rates and concentration of Cuo Nanofluid on thermal conductivity at tube side for RSM response surface (3D) plots

Fig. 14 Predicted versus actual thermal conductivity at shell side

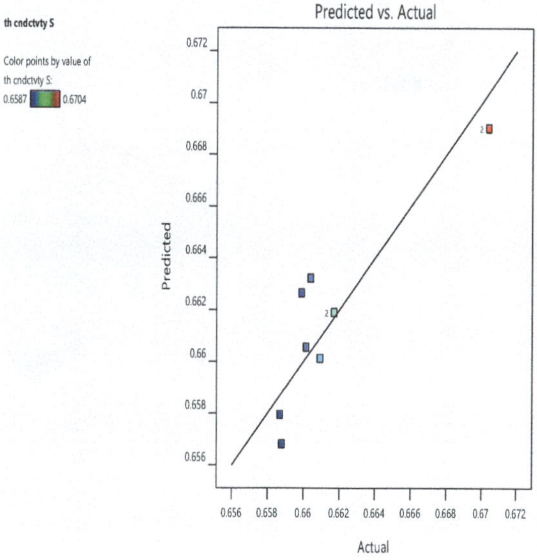

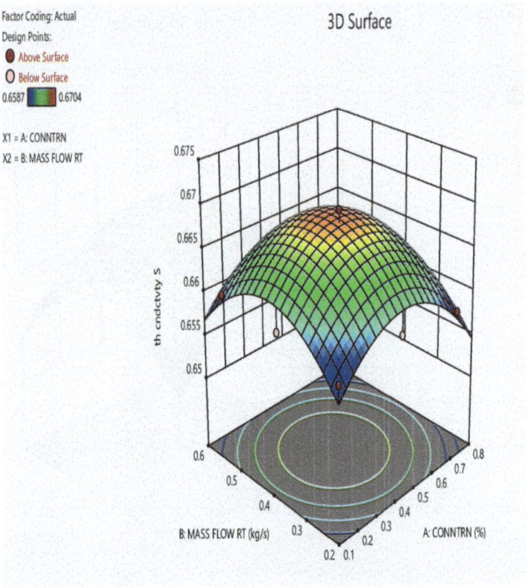

Fig. 15 Interactive effects between mass flow rates and concentration of Cuo Nanofluid on thermal conductivity at shell side for RSM response surface (3D) plots

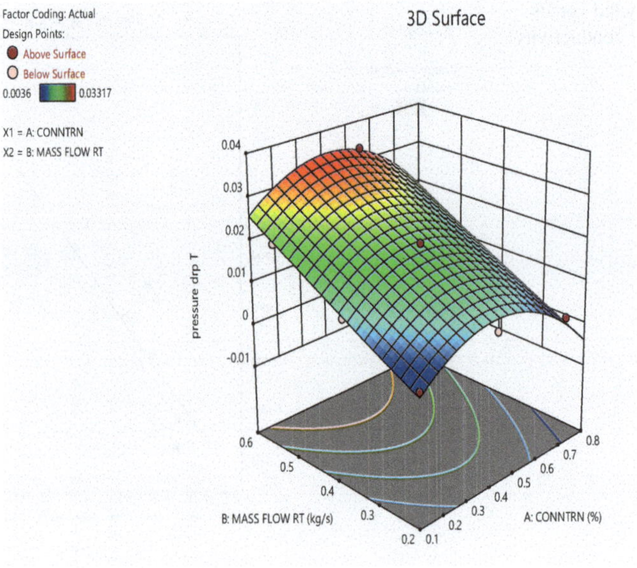

Fig. 16 Predicted versus actual pressure drop at tube side

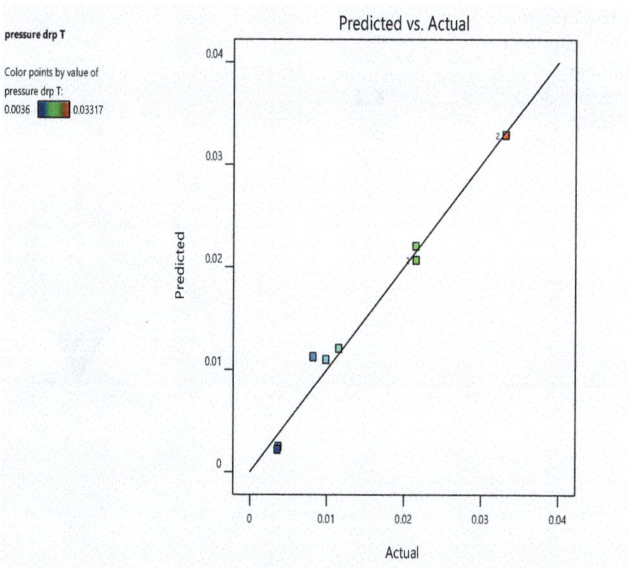

Fig. 17 Interactive effects between mass flow rates and concentration of Cuo Nanofluid on pressure drop at tube side for RSM response surface (3D) plots

Fig. 18 Predicted versus actual pressure drop at shell side

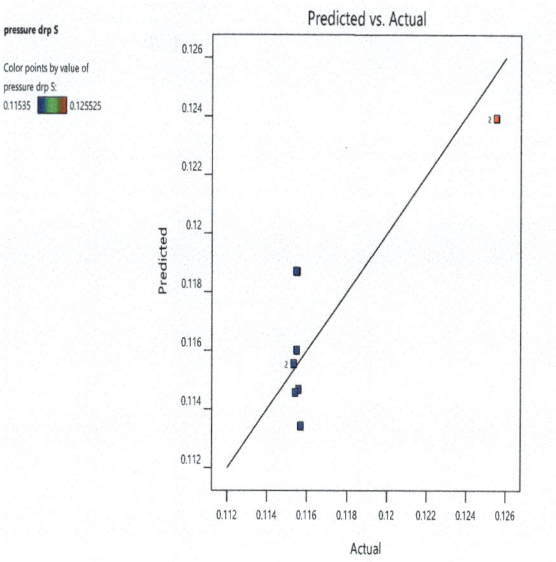

Fig. 19 Interactive effects between mass flow rates and concentration of Cuo Nanofluid on pressure drop at shell side for RSM response surface (3D) plots

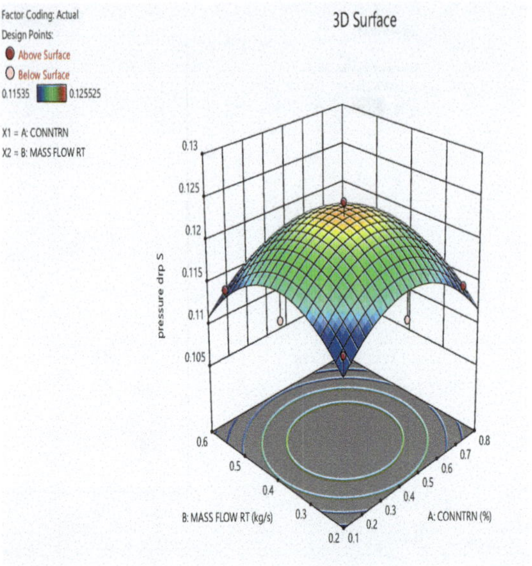

Fig. 20 Predicted versus actual fouling rate at tube side

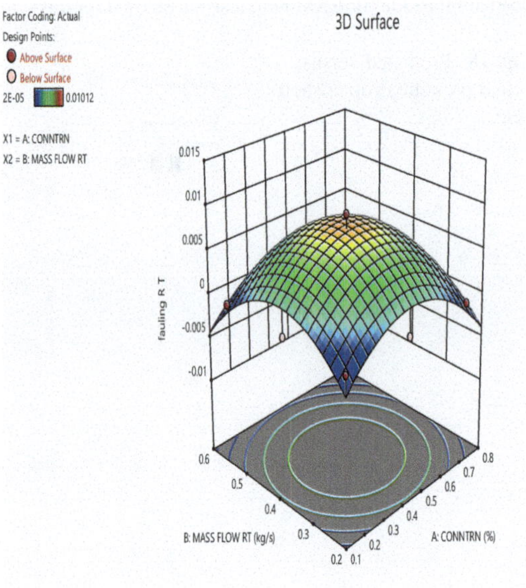

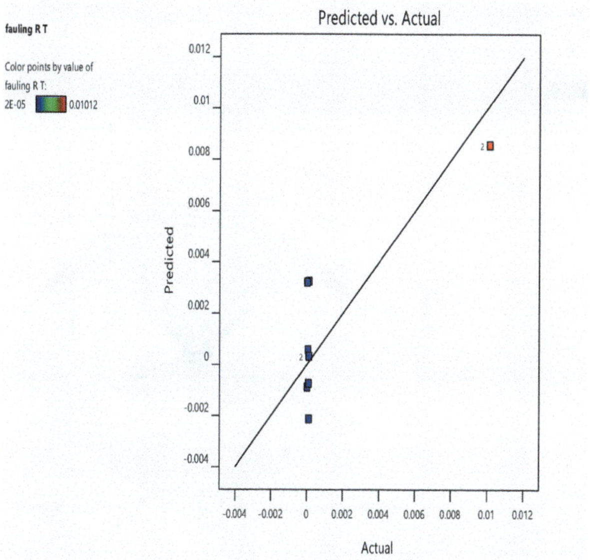

Fig. 21 Interactive effects between mass flow rates and concentration of Cuo Nanofluid on the fouling rate at tube side for RSM response surface (3D) plots

Fig. 22 Predicted versus actual fouling rate at shell side

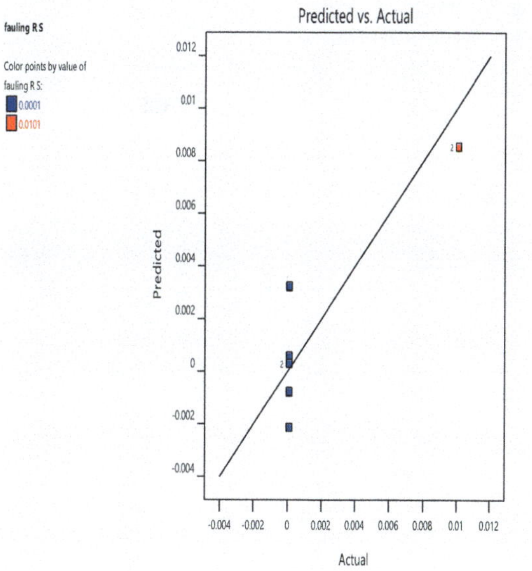

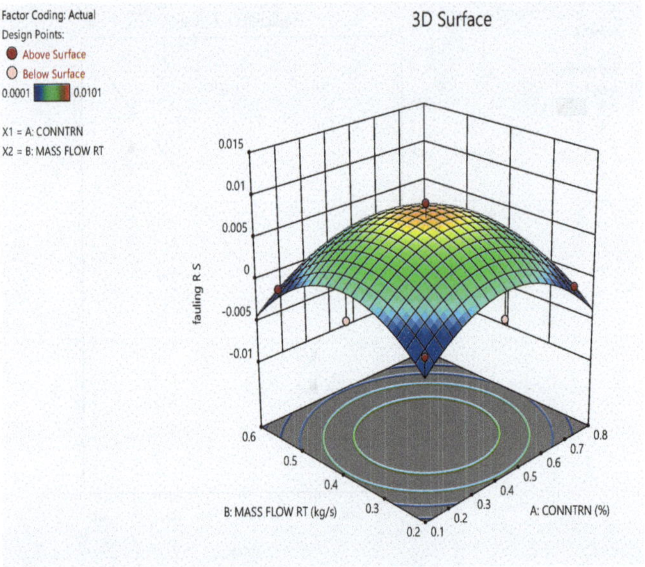

Fig. 23 Interactive effects between mass flow rates and concentration of Cuo Nanofluid on the fouling rate at shell side for RSM response surface (3D) plots

Fig. 24 Predicted versus actual heat transfer rate

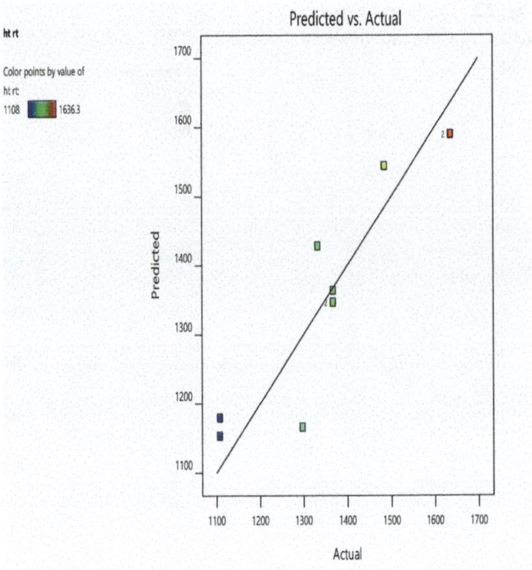

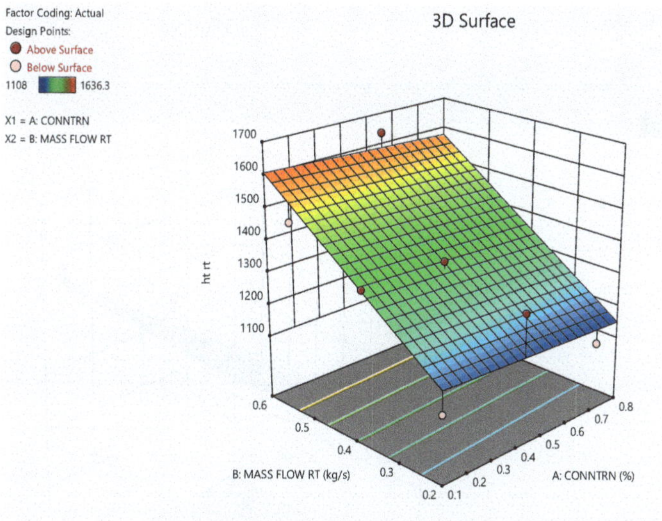

Fig. 25 Interactive effects between mass flow rates and concentration of Cuo Nanofluid on heat transfer at tube and shell side for RSM response surface (3D) plots

Fig. 26 Predicted versus actual effectiveness

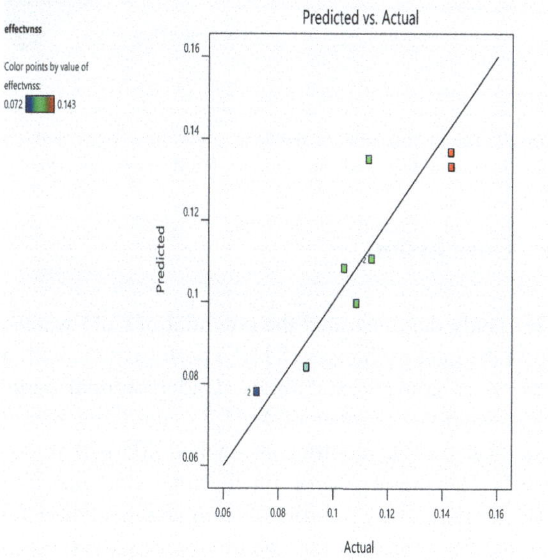

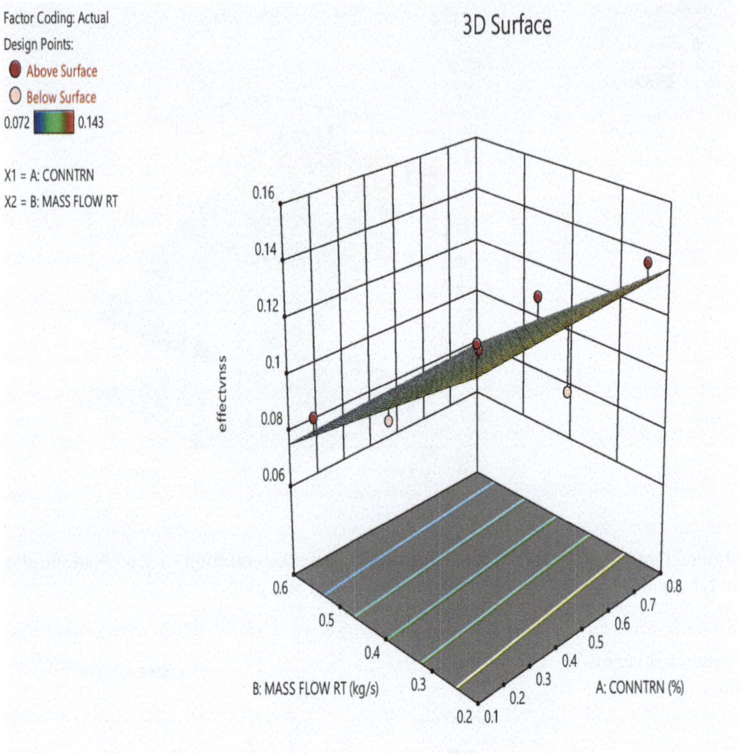

Fig. 27 Interactive effects between mass flow rates and concentration of Cuo Nanofluid on effectiveness at tube and shell side for RSM response surface (3D) plots

5 Conclusion

This study demonstrated the potential of CuO-water nanofluids in improving shell-and-tube heat exchanger (STHE) performance, with a 48% increase in heat transfer rates over conventional fluids. Under optimal conditions (mass flow rate of 0.4 kg/s, Reynolds number of 4181.75, and 0.6% nanofluid concentration), the shell-side heat transfer coefficient reached 4056.70 W/m^2·K, and the effectiveness was 0.115. The pressure drop on the tube side was 29.14 N/m^2, with the nanofluids showing excellent stability and no sedimentation for over a month. Optimization through Response Surface Methodology (RSM) revealed that nonlinear interactions influenced fouling, pressure drop, and thermal conductivity. These results support the use of CuO-water nanofluids in industrial applications, suggesting avenues for further research, including the impact of operating temperature, hybrid nanofluids, and long-term durability to enhance heat exchanger performance.

References

1. Sohrabi N, Hammoodi KA, Hammoud A, Jasim DJ, Karouei SHH, Kheyri J, Nabi H (2024) Using different geometries on the amount of heat transfer in a shell and tube heat exchanger using the finite volume method. Case Stud Therm Eng 55:10403. https://doi.org/10.1016/j.csite.2024.104037
2. Gong Q, Chulin Y, Wang W, Wang Y (2023) Experimental and numerical exploration on improved heat transfer by continuous spiral flow in the shell of spiral wound corrugated tube heat exchanger. Case Stud Therm Eng 51:103483. https://doi.org/10.1016/j.csite.2023.103483
3. Abdollahpour MS, Darbandi M (2024) Developing an algorithm to evaluate the tube and baffle choice influences on enhancing the hydrothermal performances of shell and tube heat exchangers. Case Stud Therm Eng 222:125178. https://doi.org/10.1016/j.ijheatmasstransfer.2024.125178
4. Zhe X, Ning X, Zongling Y, Ma Y, Zhao Z (2023) Binghao ZhaoDesign optimization of a shell-and-tube heat exchanger with disc-and-doughnut baffles for aero-engine using one hybrid method of NSGA II and MOPSO. Case Stud Therm Eng 41:102644. https://doi.org/10.1016/j.csite.2022.102644
5. Çolak AB, Akgul D, Mercan H, Dalkılıç AS, Wongwises S (2023) Estimation of heat transfer parameters of shell and helically coiled tube heat exchangers by machine learning. Case Stud Therm Eng 42:102713. https://doi.org/10.1016/j.csite.2023.102713
6. Li W, Kadam S, Zhibin Y (2023) Heat transfer enhancement of tubes in various shapes potentially applied to CO_2 heat exchangers in refrigeration systems: review and assessment. Int J Thermofluids 20:100511. https://doi.org/10.1016/j.ijft.2023.100511
7. Elsaid AM, Ammar M, Lashin A, Assassa GMR (2023) Performance characteristics of shell and helically coiled tube heat exchanger under different tube cross-sections, inclination angles and nanofluids. Case Stud Therm Eng 49:103239. https://doi.org/10.1016/j.csite.2023.103239
8. Ji J, Yuling Pan X, Deng JZ, Liu P (2023) Research on the heat transfer performance of an improved elastic tube bundle heat exchanger under fluid-induced vibration. Case Stud Therm Eng 146:106896. https://doi.org/10.1016/j.icheatmasstransfer.2023.106896
9. Karuppusamya S, Sambandamb P, Selvarajc M, Kaliyaperumald G, Mariadhase A, Deepake JR (2024) Enhancing heat transfer efficiency in shell-and-tube heat exchangers with SiC and CNT-infused alkaline water nanofluids. Desalination Water Treat 317:100157. https://doi.org/10.1016/j.dwt.2024.100157
10. Kelageri NK, Patil DC, Pachapuri MSA, Lingannavar R (2021) Response surface modeling for optimization of ring-stiffened polymer composite pressure hull. Compos Mech Comput Appl An Int J 12(1):75–90. https://doi.org/10.1615/CompMechComputApplIntJ.2021036491
11. Kelageri NK, Choudhary SS, Lingannavar R, Takkekar CA, Gawade SC, Modgekar PS (2024) Stability analysis and design optimization of FRP cruciform columns by response surface method. In: De A, Mukherjee PP, Pati S, Biswas A (eds) Recent trends in mechanical engineering. ICRAME 2024. Lecture notes in mechanical engineering. Springer, Singapore. https://doi.org/10.1007/978-981-97-7535-4_21

Experiential Learning in Mechatronics Through an Internship on an Autonomous Exploratory Boat Project

Archya Roy, Abhijeet Dhulekar, Faladrum Sharma, and Uday Shanker Dixit

Abstract An autonomous exploratory boat (AEB) was developed as a part of an internship. The design of the boat was oriented towards a pedagogical gadget for multi-purpose teaching. AEB can be used for learning programing through its navigation in a pond as well as aspects of environment science. It can also be presented as an example of a mechatronic product to undergraduate students. The design emphasized simplicity and affordability, making it accessible for a wide range of academic applications. Field trials were carried out and feedback was obtained from a variety of students. Field trials conducted on local water bodies showcased the boat's capability to perform scientific measurements. These were used to create visual plots and analyze environmental data. It provided a hands-on platform to learn environmental science concepts such as measuring bathymetry, water flow velocity, temperature, and total dissolved solids distribution. On one hand, the development of AEB provided a lot of learning experience for the development team. On the other hand, it is being actively used for teaching of mechatronics to a variety of students. Overall, the AEB served as an opportunity to explore aspects of mechanical design, electronics integration, control systems and contributed to science, technology, engineering, and mathematics (STEM) education. Such type of internship project also helps to develop entrepreneur skills.

A. Roy
Department of Mechanical Engineering, National Institute of Technology Durgapur, Durgapur, India

A. Dhulekar
Indian Institute of Technology Guwahati Technology Innovation and Development Foundation, Guwahati, India

F. Sharma (✉)
Centre for Indian Knowledge Systems, Indian Institute of Technology Guwahati, Guwahati, India
e-mail: fds80@iitg.ac.in

U. S. Dixit
Department of Mechanical Engineering, Indian Institute of Technology Guwahati, Guwahati, India
e-mail: uday@iitg.ac.in

A. Biswas et al. (eds.), *Recent Advancements in Artificial Intelligence and Optimization*, Lecture Notes in Mechanical Engineering,
https://doi.org/10.1007/978-981-95-9268-5_4

Keywords Flowmeter · GPS · Mechatronics education · Surface vehicle · Navigation · Sensors · STEM education

1 Introduction

One of the mandates of National Education Policy (NEP) 2020 of India is to provide hands-on skills to students (Clause 4.6 of NEP 2020) [1]. Only classroom theoretical knowledge cannot prepare industry-ready engineers. Moreover, in the era of Industry 4.0, traditional boundaries between various disciplines of engineering are diminishing. Hence, project-based learning can be more effective than traditional class-room teaching. Engineering students find an opportunity to undergo internship in an industry or research institution. If properly planned, internship can be an effective mode of imparting knowledge. This article presents the experience gained through an internship project as a case study of mechatronics education.

The first author of this article underwent a two-month summer training at Indian Institute of Technology Guwahati. Internship aimed to carry out some research in underwater technology, particularly on the design of a sensors-mounted boat for underwater exploration. There has been a lot of work on the exploration of marine environments. Exploring marine environment poses several challenges to human intervention due to high pressure, low temperature, and reduced visibility. Such constraints motivate the development of marine-based vehicles such as autonomous surface vehicle (ASV) and autonomous underwater vehicle (AUV). Key aspects in these vehicles include vehicle design, construction, navigation, control, path planning, and obstacle avoidance. There are several published articles on the design of surface and underwater vehicles. Guo and Kato [2] designed an underwater vehicle for marine education. Wang et al. [3] developed an autonomous surface vehicle (ASV) for monitoring the marine environment monitoring. The ASV was of catamaran type with two hulls. It was driven by two propellers and its overall length was 2.7 m. Beaudoin et al. [4] described some undergraduate projects of developing autonomous underwater vehicles. Some underwater gliders have also been developed for education and research [5, 6]. Khan and Ercan [7] shared their experiences of project-based education in marine technology, wherein they provided examples of developing underwater as well as surface vehicles.

Thirunavukkarasu et al. [8] designed an ASV equipped with a winch connected to the motor by a chain and sprocket system that helped to deploy an underwater robot. Carlson et al. [9] developed a portable ASV to collect bathymetry and ocean current data in hazardous environment near icebergs. The features of ASV included autonomous waypoint navigation, obstacle avoidance, and return-to-home feature. Recently, several articles have been published reporting the development of underwater and surface vehicles with a main objective of enhancing the pedagogy. The researchers focused on low cost solution [10] and training of students as well as faculty [11–13]. In fact, Rousouliotis et al. [13] used underwater vehicle for teaching Python programming.

This article describes the development of an autonomous exploratory boat (AEB) for use in science, technology, engineering and mathematics (STEM) education. It was targeted for bathymetry and water quality monitoring in ponds, lakes, and swimming pools. On one hand, this project provided a useful learning experience to the intern and his trainers; on the other hand, it became a useful teaching aid for high school students. The teaching–learning experiences gained from this project, the educational benefits, and challenges encountered throughout the process are highlighted in the subsequent sections.

2 Design and Fabrication of the Vehicle

The autonomous boat was designed and fabricated as an internship project by the first author, a third year Mechanical Engineering student of NIT Durgapur. The conceptual guidance was provided by the last author Prof. U.S. Dixit, while the second author Er. Dhulekar got involved with the intern fully during detailed design and fabrication stage. The third author concentrated on logistics and project monitoring. This methodology of proper delegation of responsibilities could complete the project in two months. It also developed a team feeling, which is an essential ingredient of a successful industry.

2.1 Modular Design and Fabrication

It was attempted to design the boat in a modular fashion. The basic module of the AEB comprised an aluminum frame, a control system and a flat-bottomed foam hull for providing stability. It was planned to fix temperature, total dissolved solids (TDS), pH, and turbidity sensors in the hull to collect data from pools, ponds, and lakes. The fabrication involved drilling, computer numerical control (CNC) milling, 3D printing, soldering, and power hacksaw. Figure 1 shows photographs of the AEB. In the basic modules, ultrasonic sensor, ping sonar, temperature sensor, and total dissolved solid (TDS) sensors have been attached. Ultrasonic sensor is suitable for short-range detection, suitable for collision avoidance, while the ping sonar is a long range sensor suitable for depth measurement. There is some space to attach other useful sensors. The major specifications are provided in Table 1. These specifications were decided based on the prior experience of the team and considering the aspects of pedagogy and cost.

The material used for the hull body was polypropylene foam, featuring a flat-bottom hull shape. An aluminum light weight frame was directly attached to the thruster, allowing the thrust force to be distributed evenly across the surface vehicle's body. Additionally, a transparent sheet was assembled with the aluminum frame to distribute the payload and control system load, thereby preventing imbalance during propulsion. The flat-bottom hull was capable of carrying up to 85 kg of payload. Two

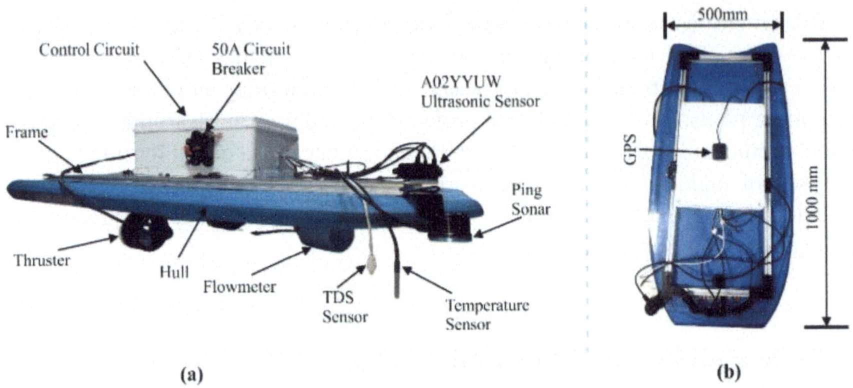

Fig. 1 Photographs of autonomous exploratory boat (AEB): **a** labelled overview and **b** top view

Table 1 Detailed specification of AEB

Specification	Parameters
Hull length	1000 mm
Width	500 mm
Weight with payload	6 kg
Battery	24 V Lithium iron, 8400 mAh
Load capacity	85 kg
Microcontroller	Teensy 4.1
Maximum speed	6 km/h

thrusters, each providing 66 N thrust, were positioned at the rear of the vehicle to propel and turn the vehicle. The microcontroller Teensy 4.1 was employed to enable autonomous navigation for collecting the bathymetric and water quality assessment data. Circuit was developed on a prototyping circuit board (PCB).

2.2 *Block Diagram Representation of Control System of AEB*

The electronic system for the Unmanned Surface Vehicle (USV) integrates a Teensy 4.1 microcontroller, serving as the central processing unit. This microcontroller manages all operations with high processing power and low latency (faster response), which is crucial for real-time decision-making. Overclocked from its normal 600 MHz to 1.008 GHz, the Teensy 4.1 enhances computing speed significantly. The block diagram of the AEB is shown in Fig. 2. It comprises sensors, actuators, microcontroller, and battery as its major constituents.

The navigation system includes a NEO-6 M global positioning system (GPS) module for accurate global positioning that communicates via a serial universal

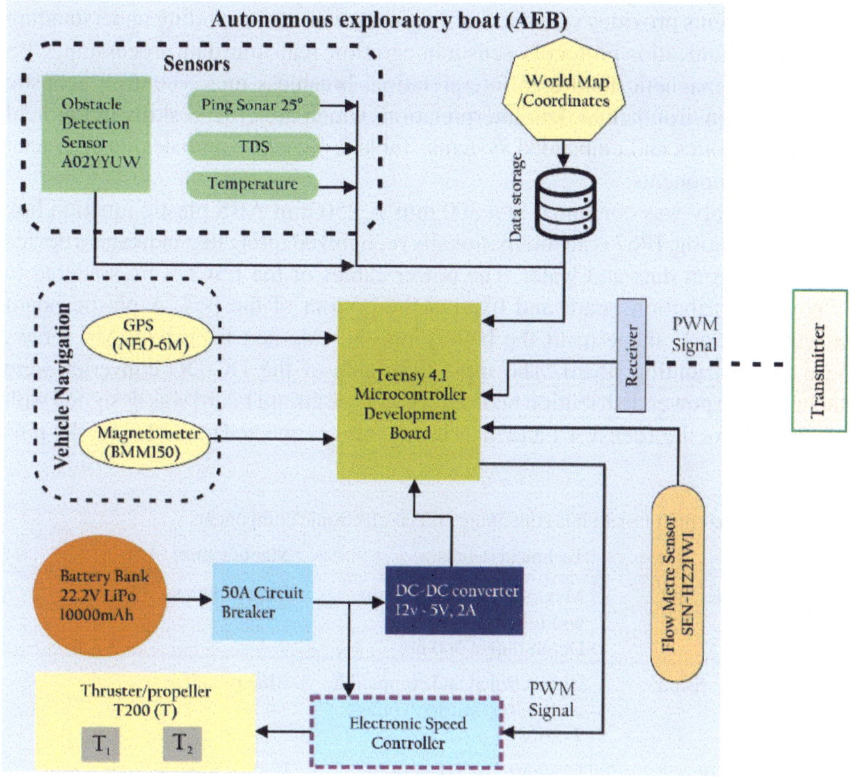

Fig. 2 Block diagram of autonomous exploratory boat (AEB)

asynchronous receiver/transmitter (UART) interface, and a BMM150 magnetometer for precise heading determination, communicating via (Interintegrated circuit I2C) protocol. I2C protocol is a serial communication protocol that uses two wires to connect devices over short distance. The propulsion system features two T200 thrusters from Blue Robotics, which are controlled using pulse width modulation (PWM) signals in electronic speed controllers (ESCs). The steering is achieved by providing different speeds to two thrusters.

Environmental sensors include a ping Sonar, using a serial UART interface with the Ping Protocol for underwater mapping, a total dissolved solid (TDS) sensor for monitoring water quality via analog signal communication, and a temperature sensor for recording water temperature variations, using single-wire bus communication. Additionally, three waterproof ultrasonic sensors are attached on the front side of the USV for obstacle avoidance.

Additional components include a power distribution board, a 8–55 V to 5 V, 2A DC-DC step-down converter, and a 50A circuit breaker connected in series with a 24 V Li-ion 8400 mAh battery. This setup enables the USV to navigate autonomously and perform environmental monitoring tasks efficiently. Working with these components provides valuable learning opportunities, including understanding various communication protocols, sensor integration, real-time data processing, GPS data parsing, magnetic field data interpretation, brushless motor control, acoustic sensing, and environmental data interpretation, which are crucial skills in the field of marine robotics and embedded systems. Table 2 presents some details of crucial electronic components.

The assembly was contained in a 300 mm × 250 mm ABS plastic junction box with an IP67 rating. IP67 is an internationally recognized rating that indicates a device is protected from dust and water. The power cables of the ESCs were soldered to the power distribution board and fixed at the bottom of the box. A plastic board was cut into an L shape to fit the battery on one side and fix it with M4 screws above the distribution board. The input terminals of the DC-DC converter were attached to the power distribution board. A prototype circuit board was designed with a cooling fan for the Teensy 4.1 and male header pins connected to the Teensy 4.1 pins

Table 2 A partial bill of materials containing crucial electronic components

Components	Technical details	Manufacturer
Ping echosounder and altimeter	Max range: 100 m, Beam width: 25°, Depth range: 300 m	Blue robotics
Basic electronic speed controller	Bidirectional and compatible with T200 thruster, voltage: 7–26 V	Blue robotics
Water flow sensor (SEN-HZ21WI)	Flowrate: 1–30 L/min Input voltage rage: 4.5–18 V	Foshan Shunde Saier sensor Co., LTD
GPS module (NEO-6 M)	Navigation update rate: 5 Hz Max Speed: 500 m/s Position accuracy: 2 m	Ublox
Magnetometer module (BMM 150)	Resolution: 0.3 μT Measuring range: ±1300 μT (x & y axis); ±2500 μT (z axis) Accuracy: 30 μT, ±2.5°	DF robot
Teensy 4.1	Microcontroller: ARM cortex-M7 at 600 MHz Memory: 7936 k flash, 1024 k RAM	PJRC
Waterproof temperature sensor (DS 18B20)	Temp. range: –55 °C to 125 °C Accuracy: 0.5 °C Voltage: −0.5 V to +6 V	Maxim Integrated Product, Inc
Total dissolved solid (TDS) sensor (SKU SEN 0244)	Measurement range: 0–1000 PPM; Measurement accuracy: ±10% Full scale (25 °C) Voltage: 3.3–5.5 V	DF robot
Waterproof ultrasonic sensor (A02YYUW)	Detecting range: 3–450 cm Operating range: 3.3–5 V	DF robot

for connecting the aforementioned components with a provision for future project expansion. This prototype circuit board, along with the GPS module, magnetometer, and DC-DC converter, was attached to the plastic board using double-sided tapes.

2.3 Navigation

An autonomous exploratory boat (AEB) for navigation utilizes the NEO-6 M GPS module to obtain the current GPS coordinates and the BMM150 magnetometer to determine the instantaneous heading direction of the boat. Initially, the algorithm sets up variables for the goal latitude, longitude, heading, heading difference, and motor throttles. It retrieves the current waypoint's coordinates and calculates the desired heading using the course to function from TinyGPSPlus open source Arduino library. The instantaneous heading direction is obtained in the range of 0° to 360°; the heading difference is computed and adjusted to fall between –180° and 180°. Based on the heading difference, the algorithm adjusts the port and starboard motor throttles to steer the vehicle towards the waypoint. The proximity to the waypoint is checked. If the vehicle is within 10 m, it proceeds to the next waypoint. This approach ensures precise and efficient navigation of the AEB through predefined routes. The flowchart of the navigation algorithm is shown in Fig. 3.

The absolute (with respect to earth) velocity of flowing is given by

$$\vec{V}_{re} - \vec{V}_{rb} = \vec{V}_{be}, \tag{1}$$

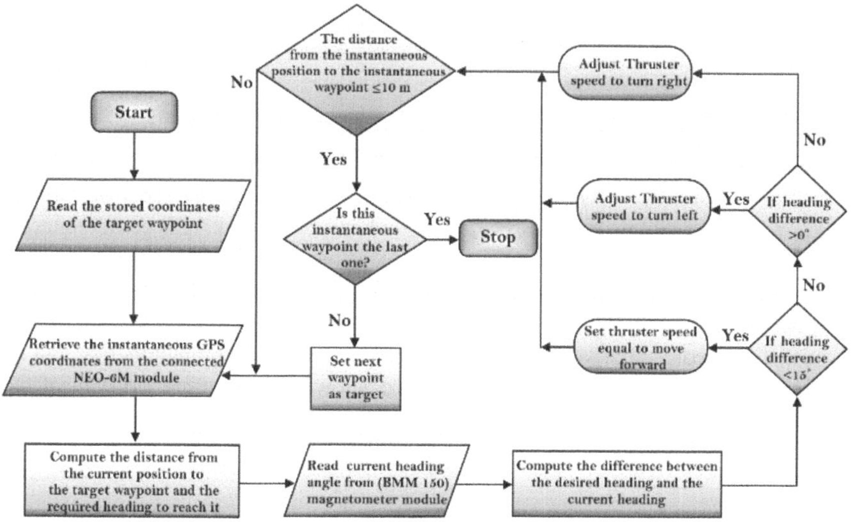

Fig. 3 Flowchart of autonomous navigation algorithm

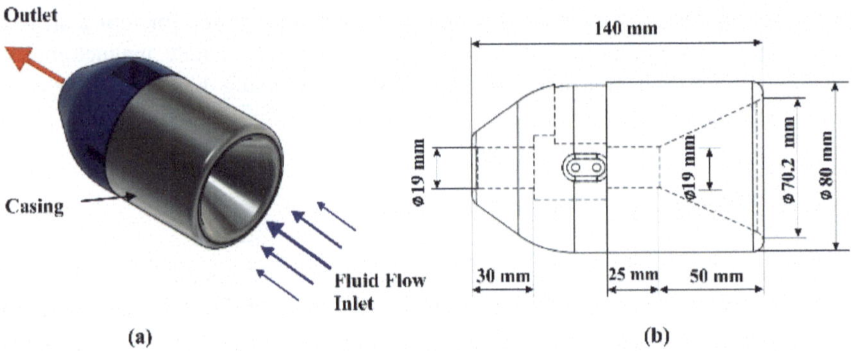

Fig. 4 Flow meter: **a** 3D model, **b** schematic diagram

where $\vec{V}_{re}$ is the river velocity in Earth frame, $\vec{V}_{be}$ is the exploratory boat velocity in Earth frame and $\vec{V}_{rb}$ is the river velocity with respect to the velocity of the surface vehicle. When boat is in still water, $\vec{V}_{re} = \vec{0}$. Thus, Eq. (1) provides

$$\vec{V}_{be} = \vec{V}_{rb}. \tag{2}$$

As $\vec{V}_{rb}$ is proportional to the flow rate Q, the magnitude of $\vec{V}_{be}$ will also be proportional to Q. To establish a relation between these two parameters, the surface vehicle was moved at different speeds in still water and the corresponding flowrates were recorded. Figure 4 shows a flowmeter that was fitted in AEB to measure Q.

3 Field Trials

Testing was carried out on Serpentine lake located in the campus of Indian Institute of Technology Guwahati, Assam, India. Later on the AEB was demonstrated at various places in the vicinity of Guwahati. Successful demonstration of AEB on various occasions established its robustness. One notable aspect is that AEB could generate interest from high school level students to professionals.

In this section, measurement of profile of lake bed, flow rate, temperature, and TDS is described. It also demonstrates the effectiveness of the GPS system. These are just some representative results. There are several other parameters that can be measured such as pH value of water. Figure 5 shows the programmed path of the AEB in Serpentine lake. The picture of the lake was obtained through Google Map.

Fig. 5 Path planning of AEB for field trial

3.1 Bathymetric Plot of Lake Bed

Data collection was carried out by the AEB navigating the lake and recording depth values in meters for various GPS coordinates at regular intervals of 10 s. These values get stored in the secure digital (SD) card. This data was then imported into MATLAB for processing and visualization. In MATLAB, values were interpolated and 3D surface plot were generated. Figure 6 shows the bathymetric plot of Serpentine Lake at IIT Guwahati.

3.2 Estimation of Velocity and Flow

Figure 7 shows the correlation between the flow rate and velocity of boat. The coefficient of correlation between Q and V_{be} is 0.9936. The following straight line was fitted:

$$V_{be} = 0.83Q + 0.11 \tag{3}$$

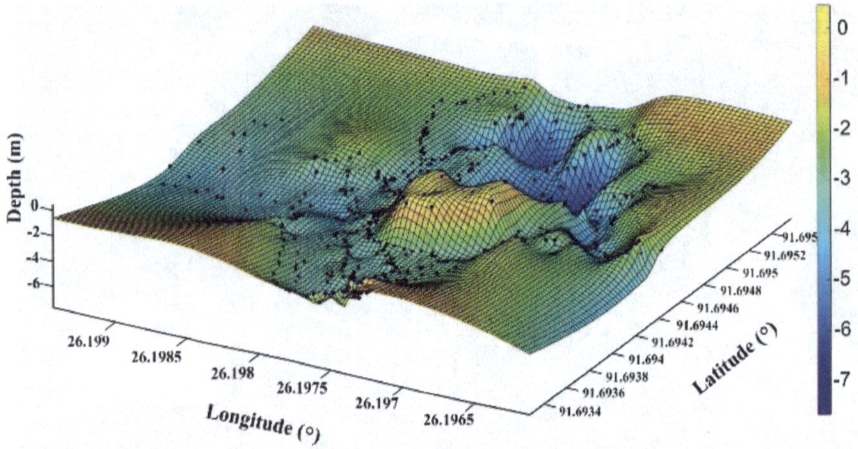

Fig. 6 Bathymetric map of the serpentine lake graph depicting depth versus longitude versus latitude

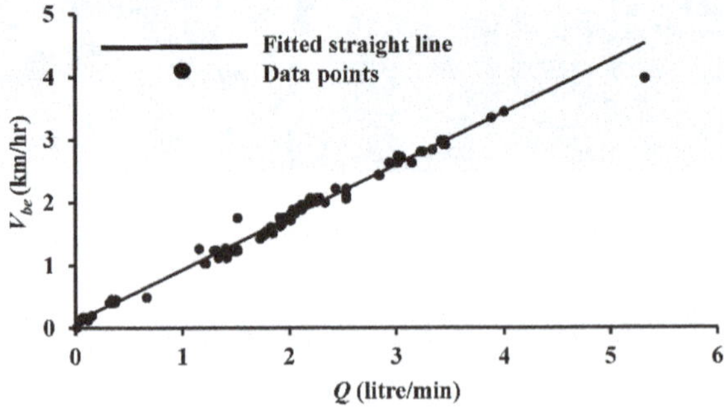

Fig. 7 Correlation between flow rate and velocity of boat

The coefficient of determination for this fitting is 0.9873. However, Eq. (3) provided a velocity of 0.11 km/hr for zero flow, which is an error. More investigation is needed to eliminate the error. Perhaps the accuracy of the flow meter needs to be enhanced.

3.3 TDS and Temperature Distribution

The results from the TDS mapping revealed distinct patterns in dissolved solids distribution (Fig. 8). This pattern suggests potential pollution sources or natural variations in water quality. The temperature mapping (Fig. 9) showed steep variations at the shallow edges of the lake, while lower variation of temperature was recorded in deeper, open water areas. There is a possibility of some noise in the data. The scope of the present project was just to show the feasibility of collecting lake data using an AEB. A detailed analysis of the collected data is part of a future study.

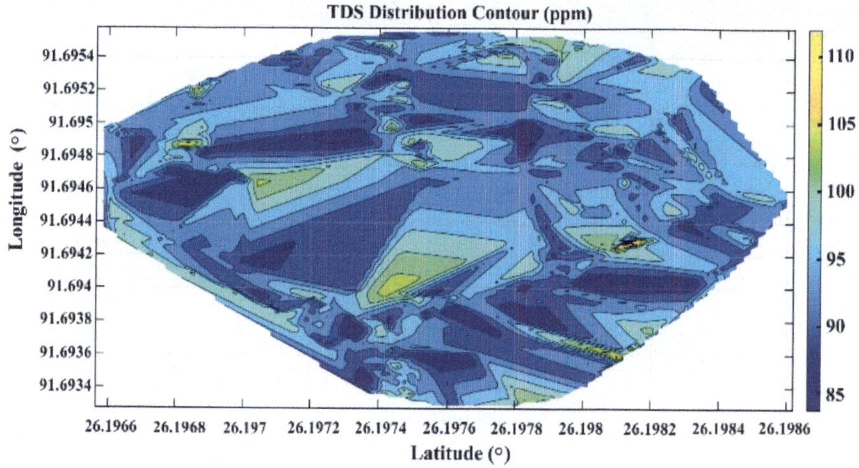

Fig. 8 Total dissolved solids at different points on predetermined path of AEB

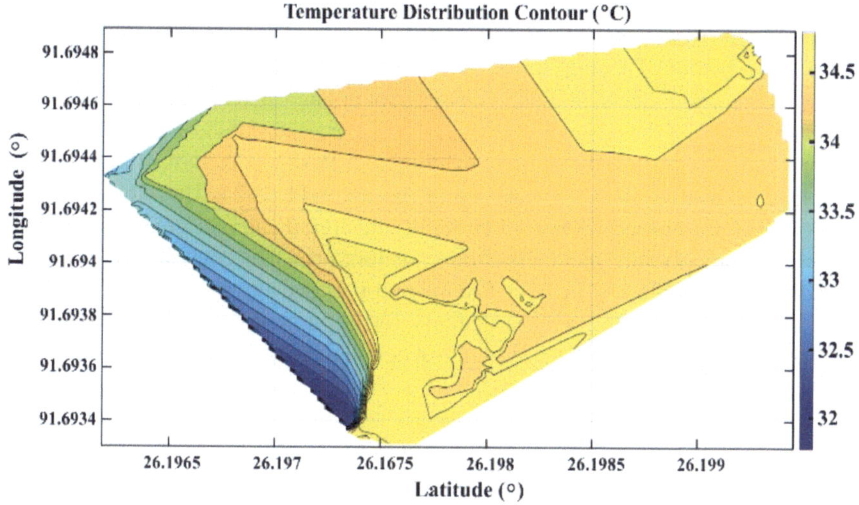

Fig. 9 Temperature distribution at different points at predetermined AEB's path

Fig. 10 The field trial of AEB in a seasonal river

3.4 Pedagogical Feedback of School and College Students

AEB was demonstrated to several class of students in order to assess its pedagogical impact. Figure 10 shows the testing of AEB in a seasonal river in Dimakuchi village of Udalguri district of Assam. It was attended by a number of rural high school and higher secondary students. Students showed a lot of interested in understanding each and every aspect of the design of AEB.

AEB was also demonstrated in an exhibition at Research & Industrial Conclave 2024 that was held at IIT Guwahati during August 9–11, 2024. The exhibition stall was visited by a number of school children, undergraduate and graduate students as well as some faculty members. Some visitors provided constructive feedback regarding its possible use in river Brahmaputra. Based on the feedback, AEB is undergoing some modification, particularly for improving its aesthetics and user-interface. It was suggested that a system can be fitted on AEB that can record as well as display the data.

4 Conclusion

AEB developed in this work provided a lot of learning experience in mechatronics to the intern student. It also provided a pedagogical gadget for STEM education. AEB has multi-purpose usage. It can be used for teaching water transportation as well as engineering science. TDS sensor can provide the data of purity of lake water. It is possible to carry out lake bed mapping. AEB can be connected with internet for online data acquisition and monitoring.

The main focus of this project was to develop a preliminary prototype based on a summer project with an objective to develop entrepreneurship skill in the intern. With the too much emphasis on theoretical and computational teaching–learning, very few students take interest in developing hardware, not to speak of a full product development. In line with NEP2020, this project provided hands-on skill and generated interest in developing a commercial product.

Acknowledgements This work is supported by IITG Technology Innovation and Development Foundation (IITG TIDF), which has been set up at IIT Guwahati as a part of National Mission on Interdisciplinary Cyber Physical Systems (NMICPS). IITG TIDF is undertaking research, development, and training activities on technologies for Under Water Exploration with the financial assistance from Department of Science and Technology, India through grant number DST/NMICPS/TIH12/IITG/ 2020. The third author expresses his gratitude to Pandit Hemchandra Goswami Foundation for supporting his stay at IIT Guwahati through the project entitled "Pandit Hemchandra Mission for popularization of Science and Assamese Culture."

References

1. National Education Policy (2020) Ministry of Education, Government of India
2. Guo C, Kato N (2008) Mini underwater glider (mug) for education. In: Workshop for Asian and Pacific Universities' Underwater Roboticians (APuuRobo 2008)
3. Wang J, Gu W, Zhu J (2009) Design of an autonomous surface vehicle used for marine environment monitoring. In: 2009 International conference on advanced computer control. IEEE, pp 405–409
4. Beaudoin L, Avanthey L, Gademer A, Vittori V, Dupessey L, Rudant JP (2012) Aquatis and Ryujin projects: first steps to remote sensing bottom of the sea by small homemade Autonomous Underwater Vehicles. In: 2012 IEEE international geoscience and remote sensing symposium. IEEE, pp 5325–5328
5. Gottschall AK (2012) Design of an underwater glider for education and research. In: 2012 IEEE/OES autonomous underwater vehicles (AUV). IEEE, pp 1–4
6. Mitchell B, Wilkening E, Mahmoudian N (2013) Developing an underwater glider for educational purposes. In: 2013 IEEE international conference on robotics and automation. IEEE, pp 3423–3428
7. Khan RR, Ercan MF (2016) Project based approach to promote interest in marine technology. In: OCEANS 2016 MTS/IEEE Monterey. IEEE, pp 1–6
8. Thirunavukkarasu G, Patrick L, Champion B, Chua L, Joordens M (2017) Design and development of a low-cost Autonomous Surface Vehicle. In: 2017 12th System of systems engineering conference (SoSE). IEEE, pp 1–6
9. Carlson DF, Fürsterling A, Vesterled L, Skovby M, Pedersen SS, Melvad C, Rysgaard S (2019) An affordable and portable autonomous surface vehicle with obstacle avoidance for coastal ocean monitoring. HardwareX 5:e00059
10. Cufí X, Figueras A, Muntaner E, Calm R, Quevedo E, Vega D, Loustau J, Gil JJ, Brito JH (2021) EDUROVs: a low cost and sustainable remotely operated vehicles educational program. Sustainability 13(15):8657
11. Bautista-Montesano R, Galluzzi R, Gomez-Aladro V, Bustamante-Bello R, Ramirez-Mendoza R (2021) Autonomous vehicles as a development platform: from high school to faculty. In: 2021 IEEE global engineering education conference (EDUCON). IEEE, pp 43–49

12. Bermúdez BY, Gutiérrez CA, Celis WC (2022) Pedagogical strategy for the generation of capabilities in the development of unmanned surface vehicles-USVs. Ship Sci Technol 16(31):19–34
13. Rousouliotis M, Vasileiou M, Manos N, Kavallieratou E (2024) Employing an underwater vehicle in education as a learning tool for Python programming. Comput Appl Eng Educ 32(1):e22693

Regularized Regressions-Based Data-Driven Modeling of BLDC Motor Characteristics for UAVs

Ushaben Patel, Anjil Shah, Purvang Mistry, Gaurav Kher, Ajay Varma, Jagat Jyoti Rath, and Ajit Kumar

Abstract Precise estimation of thrust and torque is essential for the efficient design and control of brushless DC (BLDC) motors, particularly in applications such as unmanned aerial vehicles (UAVs). The inherent nonlinear dependencies between system inputs and corresponding outputs present significant modeling challenges. In this work, a data-driven approach is adopted using regularization-based regression methods, namely LASSO, ridge, and elastic net, to predict BLDC motor thrust and torque. The investigation was carried out on a standardized test bench developed at IITRAM, Ahmedabad, where the acquired experimental data were divided into distinct training and testing subsets. The models were trained on the former and assessed on the latter to evaluate predictive performance. Among the techniques studied, the elastic net approach demonstrated the highest predictive accuracy under varying operational conditions, effectively balancing bias and variance. These findings underscore the potential of regularized regression frameworks as reliable tools for modeling BLDC motor behavior in UAVs and related electromechanical applications.

Keywords BLDC motors · Thrust prediction · Torque prediction · Regularized regression · Unmanned aerial vehicles

1 Introduction

BLDC motors, controlled by Electronic Speed Controllers (ESCs), are extensively utilized in Unmanned Aerial Vehicles (UAVs) to serve as the propulsion system, adjusting throttle inputs to generate the required thrust. Accurate modeling and prediction of BLDC motor performance, particularly thrust and torque outputs, are

---

U. Patel (✉) · A. Shah · P. Mistry · G. Kher · A. Varma · J. J. Rath · A. Kumar
Department of Mechanical and Aerospace Engineering, Institute of Infrastructure, Technology, Research and Management, Ahmedabad, India
e-mail: ushaben.patel@iitram.ac.in

A. Biswas et al. (eds.), *Recent Advancements in Artificial Intelligence and Optimization*, Lecture Notes in Mechanical Engineering,
https://doi.org/10.1007/978-981-95-9268-5_5

crucial for optimizing motor design, control strategies, and overall system performance [1]. Achieving precise propulsion modeling relies on accurately measuring the forces and moments acting on the motor shaft during flight, where the propellers are attached. However, this task is highly complex and requires the use of multiple sensors to measure both vertical and horizontal thrust components in real-time. To address this challenge, propulsion system performance is typically evaluated through static propulsion tests, dynamic wind tunnel conditions, or by employing dynamic modeling approaches.

Wind tunnel tests incur high costs, necessitate specific testing conditions [2], and may not fully capture all UAV flight stages. Conversely, static thrust stands, which measure the thrust and torque from the motor-propeller assembly, offer a cost-effective means to precisely analyze the propulsion performance of a BLDC motor-powered UAV. In [3], an experimental low-cost prototype was proposed to measure the thrust, current, and RPM of a Tarot 5008 340 KV BLDC motor with a Dragonfly 1755 propeller. Based on data collected, the thrust of the motor was modeled using a cubic regressor and efficacy for propeller parameters computation for different scenarios was presented.

The BLDC motor torque in steady state was sensed using a linear regression method with reduced set of parameters for various throttle and loading conditions on a static thrust stand in [4]. Similarly, in [5], a nonlinear regression method to identify the parameters of thrust and drag force modeled on basis of Blade Element Theory was proposed. Using experimental results on a static thrust stand and a real-time quadcopter, identification of parameters was then validated. Several other applications of static thrust stand have been explored including parameter identification of propulsion parameters in icing conditions [6], influence of vibrations on propellers [7] for multi rotors, and safe propulsion units for underground mining operations [8] among others. Several studies have focused on optimizing propulsion components such as batteries, current consumption, ESCs, and propellers [9, 10]. In [11], UAV propulsion efficiency was analyzed using different propeller and battery combinations to study payload and endurance effects on a static thrust stand. A comparison of various static propulsion thrust stands and their suitability for UAV thrust modeling was presented in [12]. However, traditional analytical and empirical models often fail to capture the complex nonlinear relationships between motor inputs (voltage, speed, load) and outputs (thrust, torque) due to factors like magnetic saturation, eddy current losses, and friction.

Recent advances in data-driven modeling have gained attention for capturing such nonlinearities using experimental data [13, 14]. Regularized regression methods like LASSO, Ridge, and Elastic Net offer better generalization and control over overfitting [15]. Beyond a few studies using SVMs, KNN, decision trees, and least-squares methods for torque modeling in BLDC motors [16], limited work has been done.

This study addresses the gap by evaluating regularized regression techniques for modeling thrust and torque in UAV BLDC motors using data from the Tyto Robotics 1585 static thrust stand. Experimental data were collected under varying voltage, speed, and load, then preprocessed and split for training and testing.

The following sections provide a detailed description of the experimental setup, data collection and pre-processing procedures, regularized regression techniques employed, and the performance evaluation criteria. The results and insights obtained from this study contribute to the development of accurate and reliable models for BLDC motor thrust and torque prediction, facilitating improved motor design, control, and optimization strategies in various applications.

2 Experimental Setup and Data Collection

The experimental setup utilizes the Tyto Robotics 1585 dynamometer, illustrated in Fig. 1, which facilitates the measurement of motor and propeller angular velocity (RPM) through an integrated optical sensor. Additionally, it enables the acquisition of thrust and torque data using calibrated load cells. The test bench's specifications, including its ability to accurately measure thrust, torque, voltage, current, RPM, resistance, and mass, along with their ranges and tolerances, are provided by Tyto Robotics [17].

Three distinct datasets, namely TDS1, TDS2, and TDS3, were created to encompass varying operational conditions using a standard BLDC motor of 920 KV and 8 inches of propeller. The TDS1 dataset was obtained for an operating range of 400 RPM (Revolutions Per Minute), while the TDS2 dataset was collected at 450 RPM, and the TDS3 dataset was acquired at 500 RPM. These datasets aimed to capture the system's performance across different rotational speed regimes, thereby enhancing the robustness and generalizability of the predictive models.

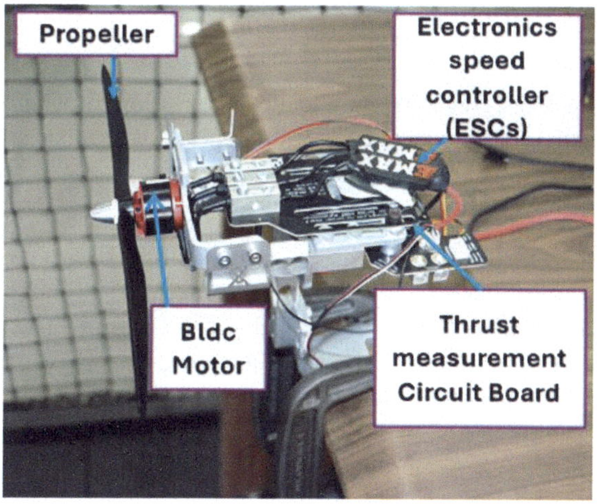

Fig. 1 Propulsion test setup. Photograph of the lab setup showing the Propeller, BLDC Motor, Electronics speed controller (ESCs), and Thrust measurement Circuit Board

3 Methodology: Regularized Regression Techniques

This study adopts regularization-based regression approaches for identifying aerodynamic characteristics of the aircraft system. In particular, it explores the application of LASSO, ridge, and elastic net models. The following sections provide a detailed examination of each technique and its implementation.

3.1 LASSO Regularization

LASSO (Least Absolute Shrinkage and Selection Operator), introduced by Tibshirani in 1996 [18], applies L1 regularization to enhance both feature selection and model simplicity. By adding an L1 penalty to the OLS objective, it drives less significant coefficients to zero, reducing overfitting and improving generalization. The regularization term:

$$\sum_{i=1}^{N}\sum_{Dj=1}(Z_i - \overline{Z}_i)_2 + \sum_{j=1}^{D}|W_j| \tag{1}$$

controls model complexity via the coefficient λ. As λ increases, more weights shrink to zero, particularly those with low influence, high variance, or negligible impact on the error. This sparsity improves interpretability and reduces multicollinearity [18].

3.2 Ridge Regularization

Ridge (L2) regularization reduces overfitting by penalizing large weights, promoting model stability and generalization [19, 20]. It modifies the loss function as:

$$\sum_{i=1}^{N} 1(y_i - \overline{y}_i) + 2\lambda\Sigma_j = 1^D W^2 j \tag{2}$$

Here λ denotes the regularization coefficient. The model optimization process involves minimizing the loss function Q (w, X), which is equivalent to solving the conditional optimization problem:

$$\min_{w} Q(w, x) = \min_{w}\left(|X\omega - y|_2^2 + \lambda|\omega|_2^2\right) \tag{3}$$

Subject to the constraint:

$$|\omega|_2^2 \leq \text{C minimize}|X_\omega - y|^2 2 \tag{4}$$

The regularization strength λ controls model complexity; larger values shrink weights, while smaller values prioritize error minimization. This helps reduce overfitting by penalizing large weights [18].

3.3 Elastic Net Regularization

Combining L1 and L2 regularizations linearly gives rise to elastic net regularization, integrating the advantages of both techniques [21]. Unlike L1 regularization, where variables can become zero, elastic net regularization allows for the establishment of conditions for a group effect with highly correlated variables.

$$\sum_{i=1}^{N}\sum_{j=1}^{D}(y_i - \overline{y}_i)^2 + \lambda 1 \Sigma_j = 1^D |W_j| \lambda 2 \sum_{j=1}^{D} \omega_j^2 \tag{5}$$

Here, λ1 and λ2 are regularization coefficients. Elastic Net is useful when many model parameters may be redundant. It is applied in areas like metric learning, cancer prediction, and portfolio optimization.

4 Results and Discussion

This study investigates the performance of LASSO, Ridge, and Elastic Net regression techniques for estimating thrust and torque of a BLDC motor using experimental data from an in-house test facility. The TDS1 dataset was structured into input-output pairs for model application.

The input dataset included PWM, accelerations ax, ay, az, voltage, current, and RPM as shown in Fig. 2, while the outputs were thrust and torque measurements as shown in Fig. 3.

To develop and validate the regression models, the dataset was divided into training and testing subsets. Specifically, 70% of the data points from the first dataset (TDS1) were randomly selected for training the regression models, while the remaining 30% of TDS1 was used for model validation and performance evaluation. In addition to using the 30% test data from TDS1, two additional datasets, TDS2 and TDS3, were employed to comprehensively assess the generalization capability of the trained models. Figures 4 and 5 show the thrust and torque validation results on the TDS1 dataset.

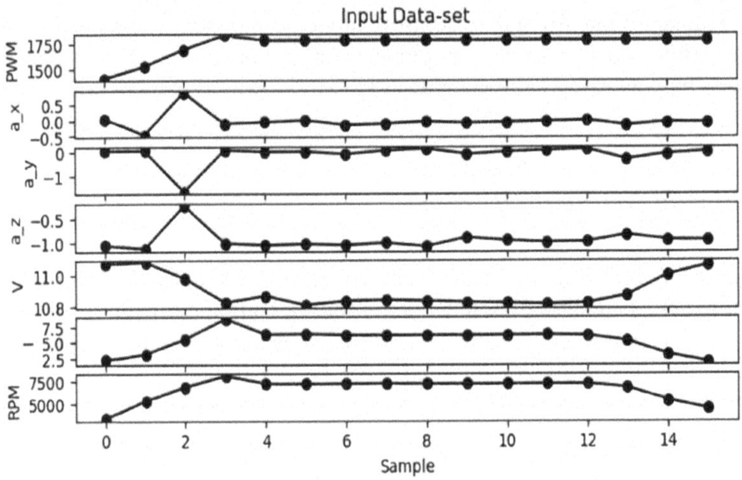

Fig. 2 Input dataset from TDS1 (PWM (pulse width modulation in μs), a_x, a_y, a_z (accelerations in m/s^2),V (voltage in volt), I (current in amp))

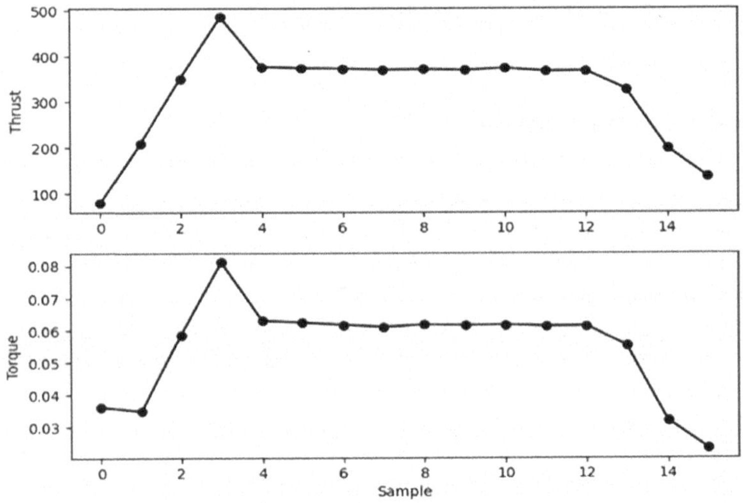

Fig. 3 Output dataset from TDS1 (thrust in gf and torque in gf-m)

Datasets TDS2 and TDS3 were used for further validation, with thrust and torque results shown in Figures 6 and 7 for TDS3. Table 1 summarizes model performance across TDS1, TDS2, and TDS3. Elastic Net achieved the best results with the highest R^2 and lowest RMSE, followed by Ridge and LASSO, while linear regression performed the worst.

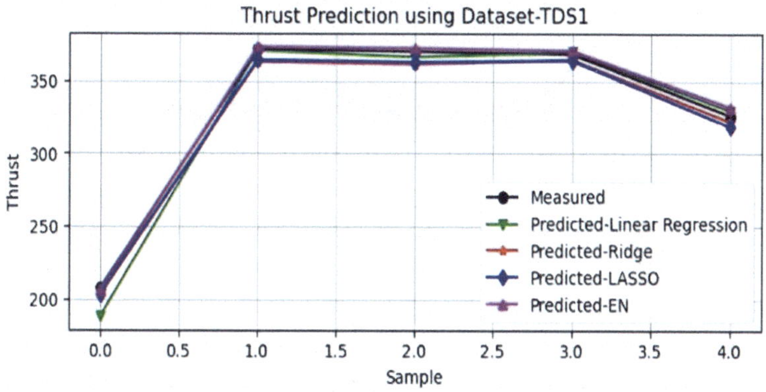

Fig. 4 Validation of thrust using Testdata from TDS1

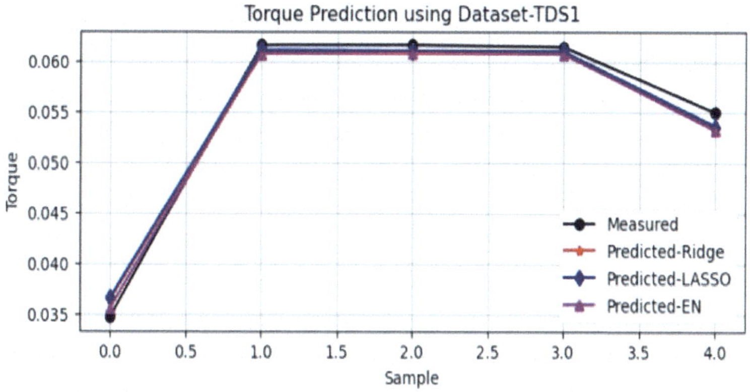

Fig. 5 Validation of torque using Testdata from TDS1

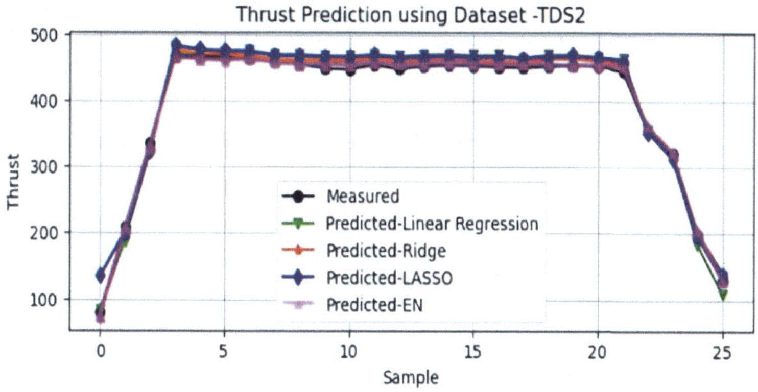

Fig. 6 Thrust prediction using Testdata TDS2

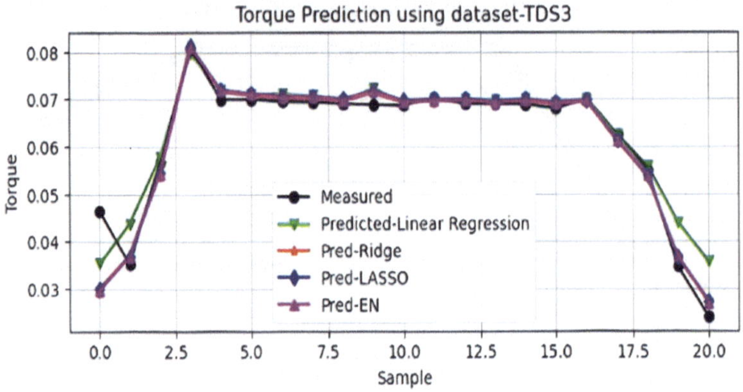

Fig. 7 Torque prediction using Testdata TDS3

Table 1 Thrust prediction

Dataset	Linear R^2	RMSE	LASSO R^2	RMSE	Ridges R^2	RMSE	Elastic Net R^2	RMSE
TDS1	0.9787	9.1171	0.9902	6.1812	0.9911	5.8821	0.9973	3.2266
TDS2	0.9814	15.190	0.9776	16.690	0.9855	13.438	0.9981	4.8979
TDS3	0.9779	16.077	0.9667	19.747	0.9715	18.27	0.9883	11.7198

Table 2 shows torque prediction results for TDS1, TDS2, and TDS3. Elastic Net performed best on TDS1, while Ridge Regression led on TDS2 and TDS3. Linear regression showed the lowest accuracy. Overall, Elastic Net and Ridge were most effective. Elastic Net gave the best thrust prediction, while Ridge was most accurate for torque. These models offer strong generalization and interpretability, making them well-suited for thrust and torque prediction in BLDC motors for UAVs.

Table 2 Torque prediction

Dataset	Linear R^2	RMSE	LASSO R^2	RMSE	Ridges R^2	RMSE	Elastic Net R2	RMSE
TDS1	0.8903	0.0034	0.9886	0.0011	0.9867	0.0012	0.9901	0.0010
TDS2	0.9539	0.0035	0.9661	0.0030	0.9736	0.0026	0.9674	0.0029
TDS3	0.8942	0.0046	0.9261	0.0039	0.9291	0.0038	0.9249	0.0039

5 Conclusions

Regularized regression techniques, particularly Elastic Net and Ridge Regression, demonstrated superior performance over traditional linear regression in thrust and torque prediction across UAV datasets. Elastic Net achieved the highest thrust prediction accuracy with R^2 values up to 0.9981 and the lowest RMSE, while Ridge Regression showed the best torque prediction on TDS2 and TDS3 with R^2 values of 0.9736 and 0.9291, respectively. These findings underscore the effectiveness of regularization in enhancing model accuracy, reducing overfitting, and supporting reliable UAV propulsion system modeling for improved design, optimization, and control.

Acknowledgment We sincerely thank the Center of Excellence in Aerospace and Defense, IITRAM, for providing the facilities and support that enabled this study.

References

1. Rath JJ, Saha S, Ikkurti HP (2012) Sliding mode scheme for speed and current control of brushless DC (BLDC) motor. In: IEEE-international conference on advances in engineering, science and management (ICAESM-2012) (pp 450-455). IEEE
2. Kumar A, Ghosh AK (2018) Data-driven method based aerodynamic parameter estimation from flight data. In: 2018 AIAA atmospheric flight mechanics conference (p 0768)
3. Kósa P, Kišev M (2022) Experimental measurement of a UAV propeller's thrust. Tehnički Vjesnik—Technical Gazette, 29(1)
4. Theys B, De Schutter J (2021) Virtual motor torque sensing for multirotor propulsion systems. IEEE Robotics and Automation Letters 6(2):4149–4155
5. Sartori D, Yu W (2019) Experimental characterization of a propulsion system for multi-rotor UAVs. J Intell Rob Syst 96:529–540
6. Løw-Hansen B, Müller NC, Coates EM, Johansen TA, Hann R (2023) Identification of an electric UAV propulsion system in icing conditions (No. 2023-01-1378). SAE Technical Paper
7. Hylmar K, Urban D, Socha V (2024) Exploring vibrations of UAV propellers: a preliminary study for characterization and fault detection. In: 2024 new trends in civil aviation (NTCA) (pp 247-251). IEEE
8. Aboelezz A, Wetz D, Lehr J, Roghanchi P, Hassanalian M (2023) Intrinsically safe drone propulsion system for underground coal mining applications: computational and experimental studies. Drones 7(1):44
9. Muzar D, Lanteigne E (2016) Experimental characterization of brushless dc motors and propellers for flight application
10. Nam H-S, Lee S, Lee H-T, Lee H-G, Lee K-J (2024) Electric propulsion system analysis and optimization for multi-rotor drones. In: 2024 IEEE aerospace conference. IEEE
11. Goli S, Kurtuluş DF, Alhems LM, Memon AM, Imran IH (2023) Experimental study on efficient propulsion system for multicopter UAV design applications. Results Eng. 20:101555
12. Wojtas M, Wyszkowski P, Mądro M, Osiewicz M, Kmita P (2023) Test stand for propellers and rotors in VTOL drone systems. Trans Aerosp Res 2023(1):67–85
13. Kumar A, Saderla S, Ghosh AK (2017) Aerodynamic parameter estimation using neuro-fuzzy model based method. In 2017 first international conference on recent advances in aerospace engineering (ICRAAE) (pp 1–5). IEEE
14. Kumar A, Ghosh AK (2019) Decision tree–and random forest–based novel unsteady aerodynamics modeling using flight data. J Aircr 56(1):403–409

15. Kumar A, Ghosh AK (2023) Regularization regression methods for aerodynamic parameter estimation from flight data. Aircr Eng Aerosp Technol 95(5):820–830
16. Hasılcı B, Mumcu TV (2022) Parameter estimation of BLDC motors by SVM for UAV propulsion systems. Erzincan Univ J Sci Technol 15(2):406–419
17. ThrustStand1580. (n.d.). Series 1580-1585. Retrieved from https://www.tytorobotics.com/pages/series-1580-1585
18. Tibshirani R (1996) Regression shrinkage and selection via the lasso. J R Stat Soc Ser B Stat Methodol 58(1):267–288
19. Hoerl AE, Kennard RW (1970) Ridge regression: Biased estimation for nonorthogonal problems. Technometrics 12(1):55–67
20. Boyko N, Kuba M, Mochurad L, Montenegro S (2019) Fractal distribution of medical data in neural network. In IDDM (pp. 307-318)
21. Zou H, Hastie T (2005) Regularization and variable selection via the elastic net. J R Stat Soc Ser B Stat Methodol 67(2):301–320

Electrochemical Machining of D2 Steel: Parameter Optimization Using KBr and $NaNO_3$ Electrolytes

Mohit Singh, P. Sudhakar Rao, and Saidaiah Kalakonda

Abstract This study investigates the influence of process parameters on the material removal rate, surface roughness, tool wear rate, and overcut in the electrochemical machining (ECM) process using two electrolytes, sodium nitrate ($NaNO_3$) and potassium bromide (KBr), at varying voltage and feed rates. A total of 19 experimental runs were conducted to explore the effects of these parameters on ECM performance. The results were analyzed using Design Expert software to develop predictive models for each output, with R^2 values calculated to assess model accuracy. The MRR model, with an R^2 value of 0.957, showed a strong correlation with process parameters, indicating that KBr, 20 V voltage, and a feed rate of 0.12 mm/min yielded the highest MRR. The surface roughness model ($R^2 = 0.876$) indicated that KBr produced higher roughness compared to $NaNO_3$. The TWR model, with an R^2 value of 0.901, revealed that $NaNO_3$caused more tool wear than KBr. The overcut model ($R^2 = 0.924$) demonstrated that lower voltages resulted in higher overcut values, with KBr exhibiting lower overcut than $NaNO_3$. Multi-response optimization using Design Expert identified the optimal parameters—20 V voltage, 0.119 mm/min feed rate, and KBr electrolyte—achieving the highest desirability score of 0.806. This research provides valuable insights into the optimization of ECM parameters to enhance machining efficiency. Further investigations could refine the models and extend the understanding of ECM process dynamics.

Keywords Electrochemical machining · D2 steel · Potassium bromide electrolyte · Sodium nitrate electrolyte · Material removal rate · Surface roughness · Tool wear rate · Overcut · Response surface methodology

---

M. Singh (✉) · P. S. Rao · S. Kalakonda
Department of Mechanical Engineering, NITTTR, Chandigarh, India

A. Biswas et al. (eds.), *Recent Advancements in Artificial Intelligence and Optimization*, Lecture Notes in Mechanical Engineering,
https://doi.org/10.1007/978-981-95-9268-5_6

1 Introduction

AISI D2 steel is one of the hard to machine cold work tool steel possessing high hardness in the range of 60–63 HR and widely used in making of extrusion dies, blanking, laminating and stamping dies, shear blades, burnishing rollers, master tools and gauges, deep drawing and forming dies thread rolling and wire dies, and other application [1]. The low machinability of D^2 steel poses significant challenges in traditional machining [2]. This has driven the advancement of non-traditional machining techniques like electrochemical machining (ECM). In this process, the material is extracted through anodic dissolution in the electrolysis. ECM eliminates tool wear, avoids reliance on the material's mechanical properties, prevents residual stress or deformation in the workpiece, and achieves a material removal rate that directly corresponds to the applied current.

Electrochemical machining (ECM) is one of the advanced machining method having high material removal rate (MRR), no thermal effects, no tool wear, no stress and good surface finish. ECM can be employed to any metal irrespective of their hardness, heat resistivity and strength, and difficulty to shape by traditional methods [3]. Many authors have reported on ECM applications for difficult-to-machine materials and have made efforts to increase MRR and surface quality (Ra), while reducing overcut (OC) and tool wear rate (TWR), and optimizing process parameters [4–6]. Neto et al. [7] investigated the ECM of SAE-XEV-F valve steel using electrolytes such as sodium chloride (NaCl) and sodium nitrate ($NaNO_3$) and optimized electrolyte flow rate, voltage, and tool feed rate to get optimum MRR, surface roughness, and overcut. Singh et al. [8] investigated the effect of inter electrode gap (IEG) on surface roughness and MRR using Al, Ti, and Cu as a workpiece under different NaCl electrolyte concentrations and confirmed that surface roughness depends on IEG. De Silva et al. [9] suggested that using lower electrolyte concentrations resulted in reduced sludge formation and significantly improved accuracy, however, the reduced conductivity led to increased heat generation within the gap. Sathish [10] investigated the effects of voltage, feed rate, current and duty ratio on MRR, tonicity, and overcut during ECM of SS (Stainless steel) 304 and observed that MRR increases with voltage. Tang et al. [11] confirmed that surface finish improves with feed and current density and NaCl electrolyte gives the highest current efficiency during ECM of S-03 stainless steel. Chakradhar and Venu Gopal [12] optimized the applied voltage, feed rate, and electrolyte concentration to achieve the best possible values for MRR, OC, and surface roughness during the ECM of EN31 steel. Mount et al. [13] observed that dissolution of SS depends on chromium content and chloride ion lowered the effect of oxide layer in ECM of SS. Tang and Yang [14] optimized multi-response parameters with different sodium-based electrolytes.

This study focuses on ECM of D2 steel with two different electrolytes, potassium bromide (KBr) and sodium nitrate ($NaNO_3$). Voltage, feed rate and electrolytes were selected as process parameters and MRR, Ra, TWR, OC as response parameters. Optimizing multiple performance characteristics is more complex than optimizing a single one. However, Response Surface Methodology (RSM) can effectively address

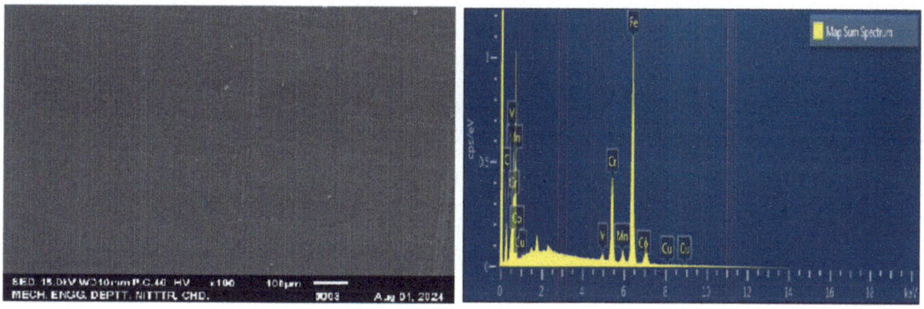

Fig. 1 SEM (100 micron) image and EDS of D2 steel

the intricate interrelationships between multiple performance characteristics; hence in this study the RSM method is used to optimize the multi-response parameter.

2 Methodology

2.1 Workpiece Material

D2 is a high-carbon, high-chromium cold-work tool steel. The specimens from stock were cut in size of 25 × 25 × 15 mm. Scanning Electron Microscopy (SEM) image and Energy Dispersive Spectroscopy (EDS) of the specimen is shown in Fig. 1.

2.2 Tool

The electrode used for machining was made of copper and had a cylindrical shape, featuring a central passage for electrolyte flow. Its exposed surface had a diameter of 18.09 mm.

2.3 Machining Setup

The setup consists of all key subsystems (Make: Metatech, Pune, India).

2.4 Process Parameter

Voltage, feed rate, and electrolyte type were selected as process parameters. Two types of electrolytes were selected for this study: one was KBr and the other was $NaNO_3$. Voltage and feed rate were continuous factors while electrolyte type is a categorical factor.

2.5 Response Parameter

MRR, Ra, TWR, OC were selected as response parameters for optimization of the process parameter. MRR and TWR is measured in gm/min, Ra in microns, and OC in mm. The average surface roughness (Ra) is quantified utilizing a Mitutoyo Surface Roughness Tester.

2.6 Experiments

A total of 24 pilot experiments were performed to determine and establish the range of process parameters. Initially, the optimal electrolyte concentrations for both electrolytes were identified, followed by the selection of the applied voltage range and feed rate range. Constant parameters during the experiment were: Temperature—30 °C, Pressure—0.50 bar, Electrode Gap—0.5 mm, Run Time—10 min, Flow rate 1.7–4 litre/min. Pilot experiments determined that a $NaNO_3$ concentration of 1.5 M and a KBr concentration of 1 M were optimal for achieving an effective MRR and subsequently the range of various parameters were also selected as shown in Table 1.

The study was conducted using response surface methodology (RSM) with a randomized I-optimal (improved response prediction accuracy and non-orthogonal) design based on the point exchange algorithm. The design is fractional factorial design having 9 model points, 5 lack of fit points, and 5 replicate points. A total of 19 experimental runs were performed with Design Expert (Version 11) software based on process parameters and their level as shown in Table 2. The quadratic model in RSM including the optimal (custom) design model is selected to capture

Table 1 Process parameters and their level

Level				
S. No	Factor	1	2	3
1	Voltage	16	18	20
2	Feed rate	0.072	0.096	0.120
3	Electrolyte type (categorical)	$NaNO_3$	–	KBr

non-linear interactions between the process parameters. No blocking was used in the design, ensuring consistency across all runs. Constant variable during the main experiments: Temperature—30o C, Pressure—0.50 bar, Electrode Gap—0.5 mm, Run Time—5 min, Flow rate 1.7–4 lit /min.

3 Results and Discussion

Table 2 shows the overall number of runs along with their associated outputs. After recording the measured response output values, the statistical analysis was performed using Design Expert software. The software also optimized the input parameters to achieve the best possible values for MRR, Ra, TWR, and overcut. Figure 2 shows all the electrochemically machined surfaces produced in each experiment. All machined surfaces with blackish appearance resulted from $NaNO_3$ electrolyte while silver white appearance resulted from KBr electrolyte.

3.1 Analysis of Material Removal Rate (MRR)

The MRR data analysis reveals a highly accurate model with a mean value of 67.63 and a standard deviation (SD) of 17.67, indicating a fairly moderate variation around the average. The coefficient of variation (C.V.) is 26.13%, showing that the variability in MRR is significant but within an acceptable range. The R^2 value of 0.9612 suggests that 96.12% of the variance in MRR can be explained by the model, reflecting a strong fit. The adjusted R^2 of 0.9418 further confirms the model's reliability after accounting for the number of predictors. The predicted R^2, at 0.8499, indicates that the model also performs well in predicting new data. Moreover, the sufficient precision value of 24.50 signifies a high Signal-to-Noise Ratio (SNR), ensuring the model's robustness and its capability to make reliable predictions. Overall, these metrics suggest a highly reliable and statistically significant model for predicting MRR. Table 3 shows the ANOVA for 2FI model for MRR.

Equations 1 and 2 describe the MRR as a function of voltage and feed rate, showing that MRR increases with increasing feed rate, while the interaction between voltage and feed rate positively influences MRR, with KBr exhibiting higher initial MRR values compared to $NaNO_3$.

For $NaNO_{3:}$

$$MRR = 800.72 - 44.15\ \text{Voltage} - 7927.17\ \text{Feed rate} \\ + 443.08\ (\text{Voltage Feed rate}) \tag{1}$$

For KBr:

Table 2 DOE with response parameter

Run	Factor 1	Factor 2	Factor 3	Response 1	Response 2	Response 3	Response 4
	A:Voltage(V)	B:Feed rate(mm/min)	C:Electrolyte type	MRR(10^{-3} g/min)	Ra(micron)	TWR(10^{-5}gm/min)	OC(mm)
1	20	0.072	KBr	100.52	1.35	57.29	1.15
2	18	0.096	$NaNO_3$	7.95	1.31	157.34	1.36
3	18	0.072	KBr	108.61	1.84	58.33	1.31
4	20	0.12	$NaNO_3$	17.64	2	156.19	0.74
5	16	0.12	KBr	140.63	1.67	59.15	0.6
6	18	0.12	$NaNO_3$	19.23	1.26	154.1	0.4
7	18	0.096	KBr	165.69	1.97	57.25	1.01
8	18	0.12	$NaNO_3$	19.14	1.26	140.09	0.36
9	18	0.072	$NaNO_3$	2.97	1.02	149.25	1.01
10	20	0.12	KBr	260.89	1.78	66.14	0.55
11	20	0.072	$NaNO_3$	2.53	0.71	152.41	1.65
12	16	0.104	$NaNO_3$	1.44	1.44	153.14	0.56
13	16	0.072	KBr	95.48	2.05	59.25	0.99
14	18	0.096	KBr	110.8	2.11	55.25	1.01
15	16	0.104	$NaNO_3$	14.08	1.33	171.17	0.69
16	20	0.096	$NaNO_3$	3.22	1.2	159.4	1.62
17	16	0.096	KBr	85	2.69	62.22	0.88
18	18	0.096	KBr	127.73	1.84	56.25	0.98
19	18	0.072	$NaNO_3$	1.42	1.04	149.28	1.11

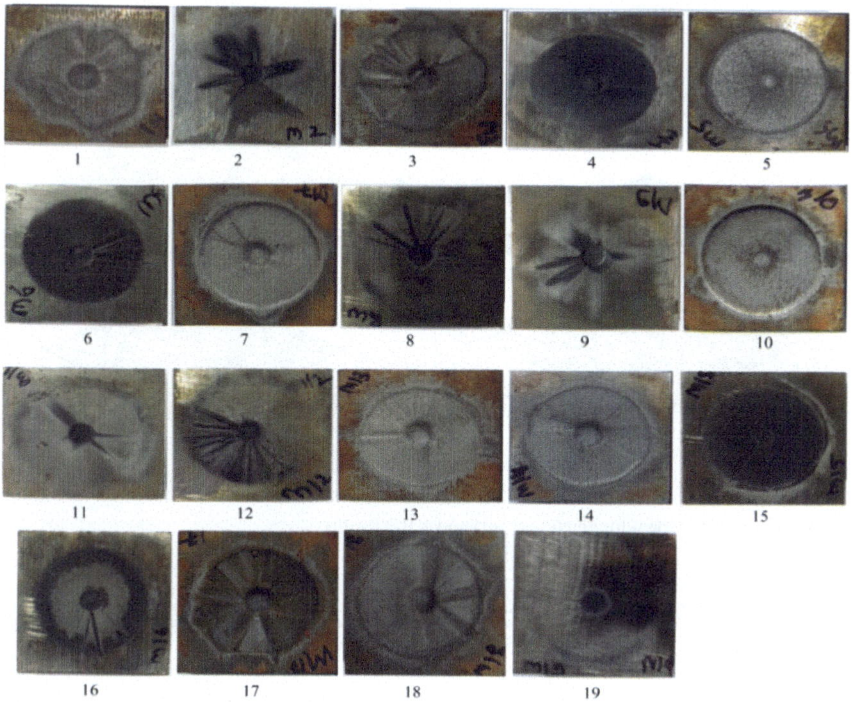

Fig. 2 All machined specimens

Table 3 ANOVA for 2FI model (MRR)

Source	Sum of squares	df	Mean square	F-value	p-value	
Model	92816.54	6	15469.42	49.53	< 0.0001	Significant
A-voltage	2637.57	1	2637.57	8.44	0.0132	
B-feed rate	6182.11	1	6182.11	19.79	0.0008	
C-electrolyte	78126.12	1	78126.12	250.14	< 0.0001	
AB	2466.49	1	2466.49	7.90	0.0157	
AC	3761.80	1	3761.80	12.04	0.0046	
BC	5610.52	1	5610.52	17.96	0.0012	
Residual	3747.93	12	312.33			
Lack of fit	2086.62	7	298.09	0.8971	0.5686	Not significant
Pure error	1661.31	5	332.26			
Cor total	96564.47	18				

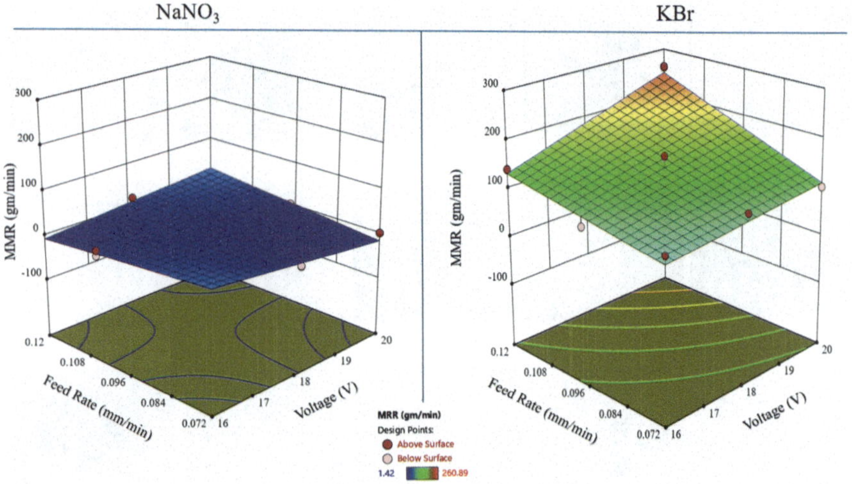

Fig. 3 3D plot of MRR with $NaNO_3$ (Left) and KBr (Right)

$$MRR = 388.08 - 24.29 \cdot \text{Voltage} - 5983.03 \cdot \text{Feed rate} + 443.08 \cdot \text{Voltage} \cdot \text{Feed rate} \tag{2}$$

The 3D plot clearly shows that machining with KBr electrolyte achieves significantly higher MRR compared to $NaNO_3$, attributed to KBr's high molar conductivity and non-passivating behavior. For both electrolytes, both feed rate and voltage have an impact on MRR, with the highest MRR observed at elevated levels of both parameters, as illustrated in the plot. The contour lines in the plot should be closer together in the regions with higher voltage and feed rate, indicating a steep increase in MRR, while they spread out more in regions with lower voltage or feed rate, indicating slower material removal. (Fig. 3).

3.2 Analysis of Surface Roughness (Ra)

The analysis of Surface Roughness (in microns) reveals key fit statistics that provide insights into the model's performance. The mean surface roughness is 1.57 microns with a SD of 0.25, suggesting a relatively low spread around the average value. The coefficient of variation (C.V.) is 15.94%, which indicates moderate variability in the data. The R^2 value of 0.8190 shows that 81.90% of the variance in surface roughness can be clarified by the model, indicating a good fit, however, the adjusted R^2 of 0.7285, while still strong, is somewhat lower, reflecting adjustments made for the number of predictors. The Predicted R^2 of 0.2063, however, is considerably lower than the Adjusted R^2, suggesting a significant discrepancy and potentially indicating issues with the model or data. A difference greater than 0.2 between these values may point to factors like model over fitting, block effects, or outliers that could affect predictive accuracy. Additionally, the adequate precision value of 10.95, well above the desirable threshold of 4, shows that the model has a good SNR and can be confidently employed to explore the design space. Despite the low predicted R^2, the model's adequate precision suggests it is capable of making meaningful predictions, though further investigation into model refinement or data quality is recommended to improve its predictive performance. Table 4 shows the ANOVA for 2FI model for surface roughness.

The contour lines in Fig. 4 for Ra are expected to be denser (indicating smaller changes in roughness) in the areas with lower voltage and feed rate, while they will spread out (indicating higher variation) at higher voltage and feed rate levels. The plot should clearly show a higher roughness in the KBr region, particularly as voltage and feed rate increases.

Equations 3 and 4 describe the Ra as a function of voltage and feed rate, showing that Ra decreases with increasing feed rate, while the interaction between voltage and feed rate positively influences Ra, with $NaNO_3$ exhibiting higher initial Ra values compared to KBr.

For $NaNO_3$

$$\begin{aligned} MRR = {} & 13.9014 - 0.6717 \cdot \text{Voltage} - 99.0873 \cdot \text{Feed Rate} \\ & + 5.5529 \cdot \text{Voltage} \cdot \text{Feed Rate} \end{aligned} \tag{3}$$

$$\begin{aligned} \text{Ra} = {} & 10.2848 - 0.5489 \cdot \text{Voltage} - 91.0053 \cdot \text{Feed Rate} \\ & + 5.5529 \cdot \text{Voltage} \cdot \text{Feed Rate} \end{aligned} \tag{4}$$

Table 4 ANOVA for 2FI model (Ra)

Source	Sum of squares	df	Mean square	F-value	p-value	
Model	3.41	6	0.5683	9.05	0.0007	Significant
A-voltage	0.2274	1	0.2274	3.62	0.0813	
B-feed rate	0.1429	1	0.1429	2.27	0.1573	
C-electrolyte	1.79	1	1.79	28.54	0.0002	
AB	0.3874	1	0.3874	6.17	0.0288	
AC	0.1438	1	0.1438	2.29	0.1561	
BC	0.0970	1	0.0970	1.54	0.2378	
Residual	0.7536	12	0.0628			
Lack of fit	0.7109	7	0.1016	11.89	0.0075	Significant
Pure error	0.0427	5	0.0085			
Cor total	4.16	18				

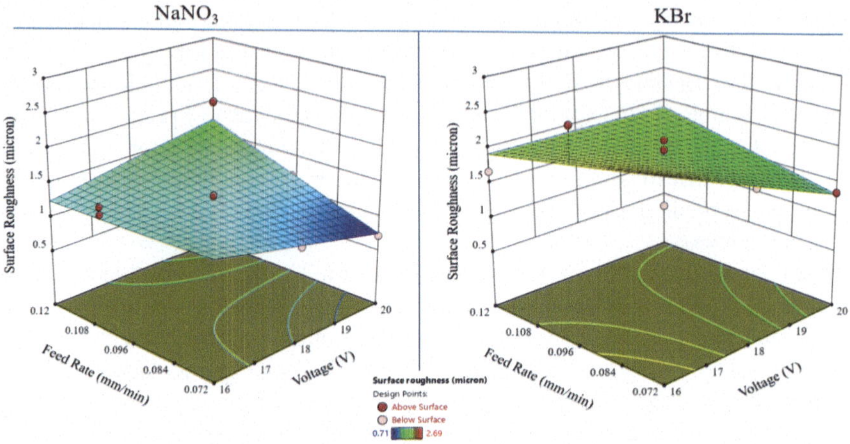

Fig. 4 3D plot of surface roughness with $NaNO_3$ (Left) and KBr (Right)

3.3 *Analysis of Tool Wear Rate (TWR)*

The statistical analysis for the TWR model yields a SD of 5.76, indicating a moderate spread around the mean value of 109.13. The R^2 value of 0.99 suggests a very high degree of correlation between the observed and predicted data points, while the adjusted R^2 of 0.99 further confirms the robustness of the model, accounting for the number of predictors used. The coefficient of variation (C.V.) is 5.28%, suggesting relatively low variability in the data relative to the mean. The predicted R^2 of 0.97 indicates strong predictive accuracy, implying that the model generalizes well to new data. Furthermore, the adequate precision value of 27.52 signifies that the model has an adequate signal-to-noise ratio, making it a reliable tool for analysis. Table 5 shows the ANOVA for quadratic model (TWR).

Equations 5 and 6 show the mathematical relationship of TWR with voltage and feed rate for $NaNO_3$ and KBr, indicating that TWR increases with feed rate and varies non-linearly with voltage, with $NaNO_3$ exhibiting higher TWR than KBr.

For $NaNO_3$

$$\begin{aligned} TWR = 778.89 - 70.72 \cdot \text{Voltage} + 492.80 \cdot \text{Feed Rate} \\ +49.07 \cdot (\text{Voltage} \cdot \text{Feed Rate}) \\ +1.80 \cdot \left(\text{Voltage}^2\right) - 7347.57 \cdot \left(\text{Feed rate}^2\right) \end{aligned} \tag{5}$$

For KBr

$$\begin{aligned} TWR = 641.37 - 68.78 \cdot \text{Voltage} + 559.27 \cdot \text{Feed Rate} \\ + 49.07 \cdot (\text{Voltage} \cdot \text{Feed Rate}) \\ + 1.80 \cdot \left(\text{Voltage}^2\right) - 7347.57 \cdot \left(\text{Feed rate}^2\right) \end{aligned} \tag{6}$$

Table 5 ANOVA for Quadratic model (TWR)

Source	Sum of squares	df	Mean square	F-value	p-value	
Model	43288.94	8	5411.12	162.96	< 0.0001	Significant
A-voltage	1.58	1	1.58	0.0475	0.8319	
B-feed rate	0.0131	1	0.0131	0.0004	0.9845	
C-electrolyte	41254.79	1	41254.79	1242.42	< 0.0001	
AB	29.80	1	29.80	0.8974	0.3658	
AC	35.03	1	35.03	1.05	0.3286	
BC	6.55	1	6.55	0.1971	0.6665	
A^2	231.37	1	231.37	6.97	0.0247	
B^2	70.95	1	70.95	2.14	0.1745	
Residual	332.05	10	33.21			
Lack of fit	69.37	5	13.87	0.2641	0.9149	Not significant
Pure error	262.68	5	52.54			
Cor total	43620.99	18				

Figure 5 illustrates the 3D surface plots of TWR for $NaNO_3$ (left) and KBr (right), showing the effect of voltage and feed rate on TWR. It is evident that $NaNO_3$ results in significantly higher TWR, while KBr maintains lower TWR across the tested voltage and feed rate ranges. The contour lines for TWR are expected to be more widely spaced in areas with lower voltage and feed rate, indicating lower tool wear. As voltage and feed rate increase, the contour lines should tighten, showing an increased wear rate. $NaNO_3$ will consistently show higher TWR across the parameter space compared to KBr.

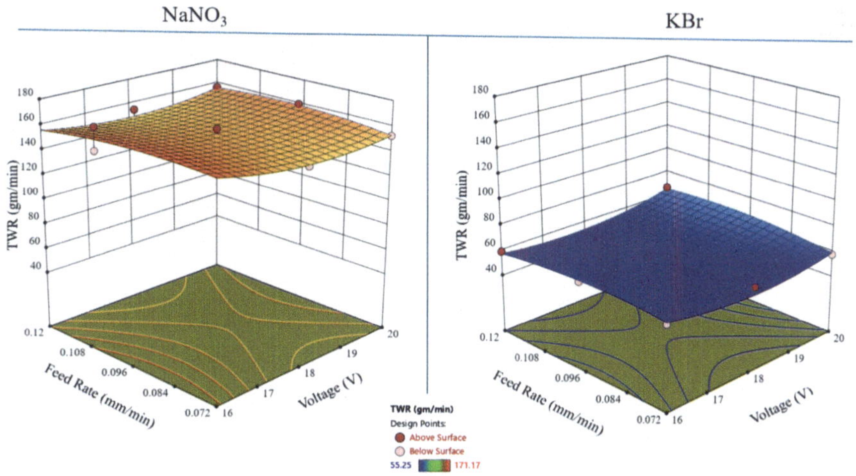

Fig. 5 3D plot of TWR with $NaNO_3$ (Left) and KBr (Right)

3.4 Analysis of Overcut (OC)

The statistical analysis for the overcut model shows a SD of 0.15, indicating a moderate spread around the mean value of 0.95. The R^2 value of 0.91 suggests a significant correlation is present between the observed and predicted data points, reflecting that the model fits the data reasonably well. The adjusted R^2 value of 0.83 indicates that, after accounting for the number of predictors, the model still maintains a good fit. The coefficient of variation (C.V.) of 16.29% suggests a higher level of variability in the data relative to the mean, indicating some dispersion in the operational cost across different conditions. The predicted R^2 value of 0.40 indicates a lower predictive accuracy, suggesting that the model may not generalize well to new data. However, the adequate precision value of 11.40 shows that the model has an acceptable SNR, making it adequate for analysis despite the lower predictive accuracy. Table 6 shows the ANOVA for Quadratic model (OC).

Equations 7 and 8 describe the Overcut (OC) as a function of voltage and feed rate, indicating that OC decreases with feed rate, while voltage has a relatively minor influence, with $NaNO_3$ exhibiting higher OC values compared to KBr.

For $NaNO_3$

$$\begin{aligned} OC = &-5.60 + 0.01 \cdot \text{Voltage} + 114.61 \cdot \text{Feed Rate} \\ &-1.46 \cdot (\text{Voltage} \cdot \text{Feed Rate}) \\ &+0.01 \cdot \left(\text{Voltage}^2\right) - 537.27 \cdot \left(\text{Feed rate}^2\right) \end{aligned} \tag{7}$$

For KBr

Table 6 Quadratic model ANOVA (OC)

Source	Sum of squares	df	Mean square	F-value	p-value	
Model	2.28	8	0.2849	11.99	0.0003	Significant
A-voltage	0.4784	1	0.4784	20.14	0.0012	
B-feed rate	1.09	1	1.09	45.79	< 0.0001	
C-electrolyte	0.0060	1	0.0060	0.2516	0.6268	
AB	0.0262	1	0.0262	1.10	0.3182	
AC	0.2510	1	0.2510	10.57	0.0087	
BC	0.0053	1	0.0053	0.2238	0.6463	
A^2	0.0058	1	0.0058	0.2421	0.6333	
B^2	0.3793	1	0.3793	15.97	0.0025	
Residual	0.2376	10	0.0238			
Lack of fit	0.2227	5	0.0445	15.00	0.0050	Significant
Pure error	0.0148	5	0.0030			
Cor total	2.52	18				

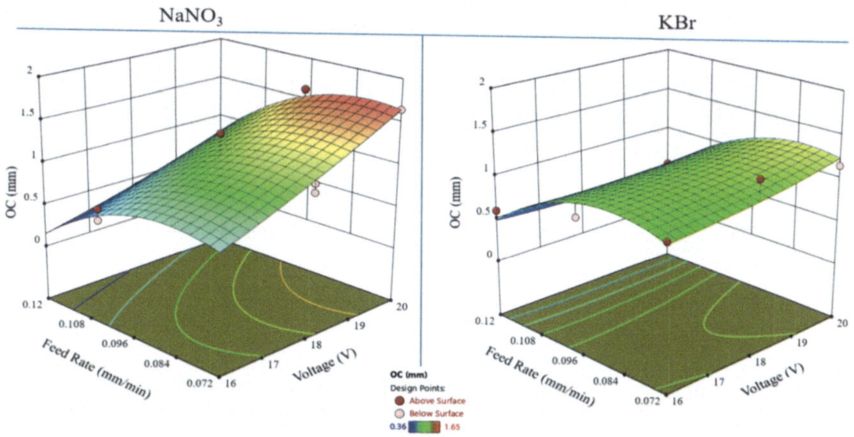

Fig. 6 3D plot of OC with $NaNO_3$ (Left) and KBr (Right)

$$\begin{aligned} OC = & -2.86 - 0.15 \cdot \text{Voltage} + 116.50 \cdot \text{Feed Rate} \\ & - 1.46 \cdot (\text{Voltage} \cdot \text{Feed Rate}) \\ & + 0.01 \cdot (\text{Voltage}^2) - 537.27 \cdot (\text{Feed Rate}^2) \end{aligned} \tag{8}$$

Figure 6 presents the 3D surface plots of overcut (OC) for $NaNO_3$ (left) and KBr (right), illustrating the influence of voltage and feed rate on OC. The results indicate that $NaNO_3$ leads to higher overcut values, particularly at elevated voltage and feed rate levels, while KBr maintains a relatively lower OC across the tested range. The contour lines for OC should be more concentrated (indicating smaller overcut) in areas of high voltage and feed rate. In regions of lower voltage and feed rate, the contour lines should be more spread out, showing larger overcut values. $NaNO_3$ will show more pronounced overcut across most of the parameter space compared to KBr.

From Fig. 7 it is clear that the input parameters—voltage, feed rate, and electrolyte Type—reveal their significant effects on both the MRR and TWR. Increasing the voltage generally leads to higher MRR, with values like 260.89 observed at 20 V, although the effect on TWR is more variable. Feed rate also shows a positive correlation with MRR, as higher feed rates, such as 0.12 mm/min, typically lead to a rise in material removal, like the jump from 140.63 to 165.69. Similarly, TWR tends to rise with feed rate, although the electrolyte type plays a key role in modulating this outcome. KBr, in particular, results in higher MRR and lower TWR compared to $NaNO_3$, indicating its superior performance in many of the runs. The combination of these factors, particularly the electrolyte type, voltage, and feed rate, suggests that optimizing these parameters can significantly improve both MRR and TWR in the process.

From Fig. 8 the Ra is influenced by the interaction of voltage, feed rate, and electrolyte type. Higher voltage generally results in a rise in surface roughness, as seen in runs where the voltage is 20 V, such as Run 10 with a roughness value of 1.78 microns. Feed rate significantly affects surface roughness, with higher rates

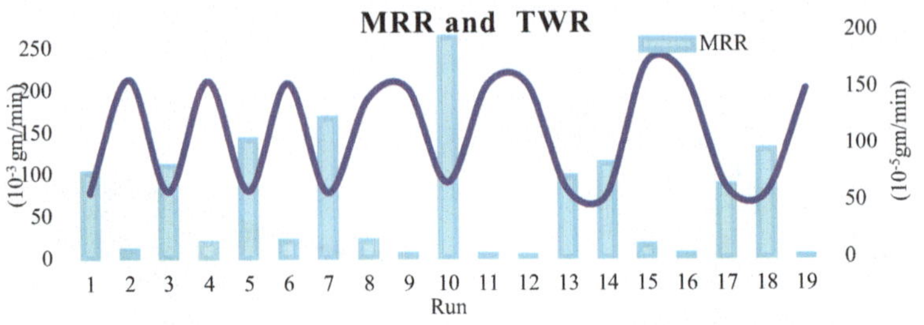

Fig. 7 Material removal rate (MRR) and tool wear rate (TWR)

often resulting in rougher surfaces, as seen in Runs 17 and 5, where the feed rate is 0.12 mm/min and surface roughness values reach 2.69 and 2.05 microns, respectively. The electrolyte type further modulates the surface finish, with KBr typically resulting in higher surface roughness values compared to $NaNO_3$, as observed in Runs 13 and 7, where KBr leads to roughness values of 2.05 and 1.97 microns, respectively. These variations highlight the complex interplay between the input parameters, suggesting that to control and minimize surface roughness, a careful balance between voltage, feed rate, and electrolyte type must be achieved.

From Fig. 8 it is clear that OC is influenced by the combined effects of voltage, feed rate, and electrolyte type. Voltage appears to have a substantial impact on OC, with higher voltage values generally leading to lower overcut values, as seen in runs with 20 V, such as Run 10, where OC is 0.55 mm. Conversely, lower voltages tend to result in higher overcut values, like in Run 2, where OC is 1.36 mm at 18 V. Feed rate is also crucial, as higher feed rates typically result in a reduction in overcut, as observed in runs with a feed rate of 0.12 mm/min (Runs 5, 6, and 10), where OC values are generally in the lower range. The electrolyte type further affects overcut, with $NaNO_3$ tending to produce higher overcut values compared to KBr, as seen in several runs. The combined effect of these factors suggests that optimizing voltage

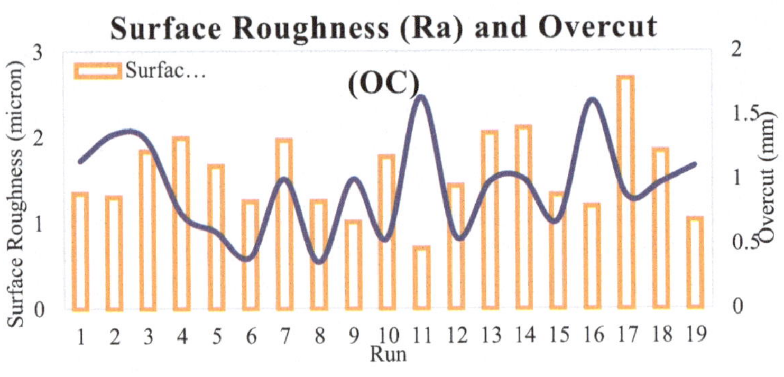

Fig. 8 Surface roughness (Ra) and overcut (OC)

and feed rate, particularly with the appropriate electrolyte, is essential for controlling and minimizing overcut in the process.

4 Conclusion

This study explored the effects of voltage and feed rate on MRR, Ra, TWR, and OC in electrochemical machining (ECM) using two electrolytes ($NaNO_3$ and KBr) and three levels of voltage and feed rate. 19 experiments were performed, and their results were examined using Design Expert software. The MRR model demonstrated a high R^2 value of 0.9612, showing that the process parameters significantly influenced MRR. The highest MRR occurred with KBr as the electrolyte, a voltage of 20 V, and a feed rate of 0.12 mm/min. The surface roughness model had an R^2 of 0.8190, with KBr producing higher roughness than $NaNO_3$. Tool wear rate (TWR) exhibited an excellent R^2 value of 0.99, indicating that $NaNO_3$ caused more tool wear than KBr. The overcut model had a moderate R^2 value of 0.91, and it was observed that lower voltages resulted in higher overcut values, with KBr leading to lower overcut compared to $NaNO_3$.

Multi-response optimization in Design Expert identified the optimal parameters as a voltage of 20 V, a feed rate of 0.119 mm/min, and KBr as the electrolyte, yielding the highest desirability value of 0.806. Also, the voltage, feed rate, and electrolyte type were found to significantly affect the ECM process outcomes. By optimizing these parameters, improvements in MRR, surface roughness, TWR, and overcut can be achieved. Further research could refine the models and enhance process efficiency.

References

1. Majhi S, Mishra T, Pradhan DM, Soni H (2014) Effect of machining parameters of AISI D2 tool steel on electro discharge machining, vol 44
2. Arsecularatne JA, Zhang LC, Montross C, Mathew P (2006) On machining of hardened AISI D2 steel with PCBN tools. J Mater Process Technol 171(2):244–252. https://doi.org/10.1016/j.jmatprotec.2005.06.079
3. Benedict GF (1989) Introduction to nontraditional machining processes. Marcel Dekker, New York, p 127
4. Xu Z, Liu J, Zhu D, Qu N, Wu X, Chen X (2015) Electrochemical machining of burn-resistant Ti40 alloy. Chinese J Aeronaut 28(4):1263–1272. https://doi.org/10.1016/j.cja.2015.05.007
5. Chen X, Xu Z, Zhu D, Fang Z, Zhu D (2016) Experimental research on electrochemical machining of titanium alloy Ti60 for a blisk. Chinese J Aeronaut 29(1):274–282. https://doi.org/10.1016/j.cja.2015.09.010
6. Zhu D, Yu L, Zhang R (2018) Dissolution effects with different microstructures of inconel 718 on surface integrity in electrochemical machining. J Electrochem Soc 165(16):E872–E878. https://doi.org/10.1149/2.0761816jes
7. da JCS N, da Silva EM, da Silva MB (2006) Intervening variables in electrochemical machining. J Mater Process Technol 179(1–3):92–96. https://doi.org/10.1016/j.jmatprotec.2006.03.105

8. Singh G, Rao PS, Singh R (2023) Effect of inter-electrode gap on surface characteristics in conventional electrochemical machining. Mater Today Proc:13–17. https://doi.org/10.1016/j.matpr.2023.10.112
9. De Silva AKM, Altena HSJ, McGeough JA (2003) Influence of electrolyte concentration on copying accuracy of precision-ECM. CIRP Ann—Manuf Technol 52(1):165–168. https://doi.org/10.1016/S0007-8506(07)60556-3
10. Sathish T (2019) Experimental investigation of machined hole and optimization of machining parameters using electrochemical machining. J Mater Res Technol 8(5):4354–4363. https://doi.org/10.1016/j.jmrt.2019.07.046
11. Tang L, Li B, Yang S, Duan Q, Kang B (2014) The effect of electrolyte current density on the electrochemical machining S-03 material. Int J Adv Manuf Technol 71(9–12):1825–1833. https://doi.org/10.1007/s00170-014-5617-x
12. Chakradhar D (2011) Multi-objective optimization of electrochemical machining of en31 steel by grey relational analysis. Int J Model Optim 1(2):113–117. https://doi.org/10.7763/ijmo.2011.v1.20
13. Mount AR, Howarth PS, Clifton D (2003) The electrochemical machining characteristics of stainless steels. J Electrochem Soc 150(3):D63. https://doi.org/10.1149/1.1545463
14. Tang L, Yang S (2013) Experimental investigation on the electrochemical machining of 00Cr12Ni9Mo4Cu2 material and multi-objective parameters optimization. Int J Adv Manuf Technol 67(9–12):2909–2916. https://doi.org/10.1007/s00170-012-4703-1

Hydrophobic Porous Polymeric Surface to Enhance Nucleation Boiling Heat Transfer: A Road to Achieve 1000 kW/ m^2 Heat Removal from Microelectronics

Ravi Pippal, Pushpendra Kumar Shukla, and Sumit Sinha-Ray

Abstract The study illustrates a simple yet cost-effective measure to enhance boiling heat transfer characteristics of de-ionized (DI) water by employing a polymeric nanofibrous texture on a superheated substrate mimicking high-performance microelectronics. The nanofibers were coated on the copper substrate as a base, using a novel method, namely supersonic solution by blowing for 90 s. The results revealed that the onset of nucleate boiling (ONB) for the nanotextured surface was 3.23 °C, whereas for bare copper surface it was 9.07 °C. Additionally, the maximum heat flux in the nucleate boiling region on the nanotextured surfaces (1265 kW/m^2) was enhanced by 143% compared to the bare copper surface (517.1 kW/m^2). Moreover, the boiling heat transfer coefficient (BHTC) for the coated surface was approximately 5.1 times better than the bare surface. With introduction of the nanofluid (a suspension of CNT in water), an early ONB (3.2 °C) on the bare surface was seen, mostly due to thin fouling layer. With nanotextured and nanofluid combination, the maximum heat flux was 106% better than bare and water combination. This is to be noted that all the experiments were tested between a supply voltage of 40–140 V. The nanofiber coated sample exhibited hydrophobic behavior with contact angle of 137.7°. The higher contact angle on the coated surface facilitated earlier ONB. The presence of ultrafine nanofibers on the copper surface increased the bubble departure frequency due to cage-like structure, which chopped the growing bubbles into smaller bubbles. The nanofibers also overheated the surrounding fluid of vapor bubbles, which impeded the growth and kept the bubbles smaller. Both phenomena, aided by the higher contact angle of fibrous network improved the overall nucleation boiling heat transfer.

Keywords Pool boiling · Nanotexture · Nucleate boiling · Hydrophobicity · Microelectronics

R. Pippal · P. K. Shukla · S. Sinha-Ray (✉)
Department of Textile and Fibre Engineering, Indian Institute of Technology-Delhi, Hauz Khas, New Delhi, India
e-mail: ssinharay@textile.iitd.ac.in

A. Biswas et al. (eds.), *Recent Advancements in Artificial Intelligence and Optimization*, Lecture Notes in Mechanical Engineering,
https://doi.org/10.1007/978-981-95-9268-5_7

1 Introduction

In recent decades, the proliferation of compact computing hardware has been remarkable, driven by the widespread adoption of the internet. Advancements in nanoelectronics have facilitated the miniaturization of electronic devices, but these developments face significant challenges due to the demanding thermal management requirements. For example, the operating temperature of silicon chips must not exceed 70 °C. Above this temperature range the efficacy of chips decreased by 10% for every 2 °C increments in temperature [1]. The intensified heat flux generated by microelectronics is exacerbated by the challenges associated with effective thermal management. Forced convection by air is not capable enough to achieve the necessary cooling requirements in a given space at a low noise level [2]. To achieve these necessary cooling requirements, pool boiling is an attractive solution due to its high heat removal capacity in terms of latent heat removal [3].

Dhir [4] provides a thorough exploration of the fundamental physical and engineering principles of the pool boiling for pure liquids. In addition to that, this theoretical review offers significant technological developments made in this field up to the year 1998. The underlying phenomena and engineering principles governing heat transfer on nucleate pool boiling are described in reference [5]. The alteration of the surface characteristics can lead to substantial improvements in the performance of boiling heat transfer [6]. Moreover, there are several other factors that can affect nucleate boiling, resulting in significantly enhanced critical heat flux (CHF). The improvement in nucleate boiling heat transfer (NBHT) performance is fundamentally governed by two key factors: nucleation site density and frequency of the departing bubble from the heated surface [6]. The addition of micro/nano texture on the heated surface is expected to lead to substantial enhancement in the performance of boiling heat transfer [6]. This improvement is primarily attributed to the increment in nucleation site density, which facilitates the initiation and growth of bubbles, thereby augmenting the overall heat transfer phenomenon. The wettability of a surface is a key factor that influences the performance of pool boiling heat transfer. Several researchers have investigated the application bi-philic surface texturing for heat transfer enhancement. The bi-philic surface can be defined as the formation of a hydrophobic texture on the hydrophilic surface [7–9]. The incorporation of superhydrophobic spots on a superhydrophilic surface has shown substantial improvement in boiling heat transfer performance [7–9].

Electrospinning, spraying, and electroplating, etc., are low-cost, straightforward, and scalable techniques to fabricate the micro/nanotextured surface [10–15]. These techniques can be used as per requirement and can be scalable for commercial applications. The pool boiling of water and ethanol on a copper-plated electrospun nanofiber-coated surface revealed a significant enhancement in the heat transfer as compared to boiling on the bare surface [10]. Significant performance enhancements have been observed in boiling when copper-plated nanofibers are deposited on a copper substrate via supersonic blown for such fluids, i.e., Nove 7300, ethanol, and water-n-heptanol [11, 12]. The pool boiling of Novec 7300 on a polymeric nanofibrous surface via supersonic blowing revealed significant enhancements in boiling performance [13]. These findings suggest that the boiling performance is influenced by molecular interactions between fluid and heated surfaces.

An alternative approach to improve the boiling heat transfer, incorporates nanoparticles into the base fluid, resulting in a nanofluid, which has been extensively researched [16, 17]. While nanofluids have the tendency to either increase or decrease the NBHT, generally, it has been found to show a significant augmentation in the CHF [16, 17]. These effects can be attributed to changes in surface characteristics, as well as a substantial improvement in thermal conductivity and reduction in the surface tension resulting in in the enhancement in NBHT [16, 17]. However, it is important to note that the observed increase in thermal conductivity of nanofluids is not anomalous and can be explained by the classical effective medium theory.

2 Literature Review and Objectives

Numerous studies have been conducted to investigate the effect of modification in surface properties such as wettability, roughness, and micro and nano-structure on augmentation on CHF, HTC, and vapor bubble departure frequency and suppression in vapor bubble diameter [18–22]. Consideration of capillary length scales in the system of 1D architecture on the order of 100 nm can enhance the HTC. The surface superheat can be significantly reduced by the combined effect of capillarity, enhanced surface area, and vapor trapping [18]. The augmentation of nucleate boiling has drawn attention to polymer nanofibers in recent years. In addition to a vapor bubble cage and wicking effect, nanotexture enhances the net boiling surface area. To enhance the cooling efficiency, a polymer nanofibre coated surface can suppress and reverse the liedenfrost phenomenon due to the capillary wicking via membrane pores [23]. The pool boiling of ethanol and water on a copper surface coated with copper-plated electrospun nanofibers was studied by Jun et al. [10]. A notable enhancement in BHTC was observed using a copper-plated nanotextured surface, as the fluid around the bubbles emerging through the nanotextured surface had a higher average temperature, leading to an earlier onset of nucleate boiling (ONB). However, surface delamination affects the nanotexturing on the surface. In order to resolve the delamination issue mentioned in [11], Sahu et al. [12] studied the pool boiling of water–ethanol combination, Novec 7300, and self-rewetting fluid on the flattened copper surface coated via electro-assisted supersonic blowing to fabricate ultrathin nanofibers. For the same surface superheat, they have reported heat flux 2–7 times higher on the copper-plated nanofiber coated surface than on the bare copper surface. These nanofibers chopped the surrounding vapor bubbles into smaller bubbles by acting as cage. Sinha-Ray et al. [13] described a similar phenomenon wherein pure polymeric ultrathin nanofibers exhibited at-par performance of metal plated nanofibers; hence it eliminated the need of an expensive electroplating system.

To modify the surface, coating of polymer ultrafine nanofibers can be scalable, inexpensive, and user friendly. Apart from the surface modification techniques, pool boiling phenomena were also studied for various kinds of nanofluids. The suspension/dispersion of nanofluids in some base fluids can enhance the thermo-physical properties of fluids such as thermal conductivity and it can also change the behavior of boiling, with a downside of possible fouling effect. To study the pool boiling characteristics, variety of nanofluids containing distinct nanoparticles such as-CuO, Fe_2O_3, Al_2O_3, ZrO_2, and SiO_2 were utilized [24–28]. Most of these findings showed similar results of improvement in CHF and a decline in HTC due to deposition of the nanoparticles on the surface. Apart from these nanoparticles carbon-based materials (single wall CNTs) can also be used for heat transfer enhancement [29]. SWCNT improves the pool boiling heat transfer by enhancing the CHF and BHTC and along with that it reduces the wall superheat. The size of nanoparticles, nanofluid concentration, working fluid, and surface characteristics are some the factors that affect heat transfer performance [30–32].

According to the present literature review, early ONB is reported from 1D nanotexturing on the heated surface. On the other hand, nanofluids enhanced the CHF but it degrades the surface as well. In the case of pure polymer nanofiber coating, the texture was considered to be robust during the pool boiling of Novec 7300 [13]. However, the authors from same experiment claimed a maximum heat flux of about 300 kW/m^2 for Novec 7300, and as high as 1000 kW/m^2 for DI water. However, for DI water, the nanotexture completely disappeared from the surface and they obtained a pool boiling behavior almost identical to that of a bare surface. As per the best of knowledge of authors, no significant studies have been undertaken using a combination of the nanofiber textured surface and nanofluid, especially with SWCNT, a material with one of the highest reported thermal conductivities. The aim of this work is to harvest the benefits of both ultrafine nanofibers and nanofluid for HTC augmentation, with any possible improvement in nucleate boiling. Pool boiling of DI Water and SWCNT/DI water-based nanofluid is experimentally studied on bare copper, coated surface with supersonic solution blown nanofibers, and coated surface with polymerous nanofibers along with MWCNT via supersonic solution blown nanofibers.

3 Materials and Method

3.1 Materials

Poly(vinylidene fluoride) (PVDF) was acquired from Thermo Fisher Scientific (Alfa Aesar). Dimethylformamide (DMF) was acquired from Alfa Aesar. Acetone was purchased from Thermo Fisher Scientific. De-ionized water (DI Water) was acquired from Organo Biotech Laboratories Pvt Ltd. Single-walled carbon nanotube (SWCNT) dispersion (2 wt% in water) was obtained from Nanoshel Inc. Aluminum block, copper cylinder, and Teflon were acquired from local suppliers. Thermal paste (MX-4 Thermal Compound) was obtained locally. Thermometer (HH374) and K-type thermocouple were procured from Omega Engineering. Cartridge heaters and immersion heater (rated 100 W each) were sourced from a local manufacturer in Delhi, India.

3.2 Nanofibers' Coating on Copper Substrate via SSB Technique

The schematic of experimental set-up is shown in Fig. 1a. An 8 wt% of PVDF polymer solution was prepared in DMF and acetone as the solvent in 1:1 ratio. The polymer solution was kept on a hot plate magnetic stirrer at 60 °C and 600 RPM overnight to ensure homogeneous mixing. The prepared polymer solution was loaded into a 5 ml syringe and pumped through a 22G needle at a controlled flow rate of 0.7 ml/hr using a syringe pump (New Era Pump system InC.). A compressor, functioning at150 psi, was linked to a supersonic de Laval nozzle (Silvent 209 L). The needle tip, dispensing the polymer solution, and gas nozzle exit were positioned 1.5 and 1 cm apart in the horizontal and vertical direction, respectively. The nozzle was grounded, while a 6.5 kV electric potential was connected to the needle tip. The nanofibers were collected on a copper block positioned 25 cm from the nozzle exit. The nanofibers were deposited on a copper substrate for a duration of 90s.

3.3 Preparation of Nanofluid

To prepare the nanofluid suspension, 0.01 wt% of SWCNT dispersion (2 wt% original) was prepared in DI Water, by diluting an adequate mass of original SWCNT suspension in 100 gm DI Water. The suspension of SWCNT and DI Water was kept in an ultrasonic bath for 30 min for proper mixing.

3.4 Characterization

Subsequent to the coating process, scanning electron microscopy (SEM) and contact angle measurements were done to characterize the surface properties of both bare and polyvinylidene fluoride (PVDF)-coated copper substrates. Contact angle measurements indicated that the bare copper surface exhibited hydrophilic behavior, with a contact angle of 85.6°. In contrast, the nanofibrous PVDF-coated surface demonstrated hydrophobic properties, characterized by a contact angle of approximately

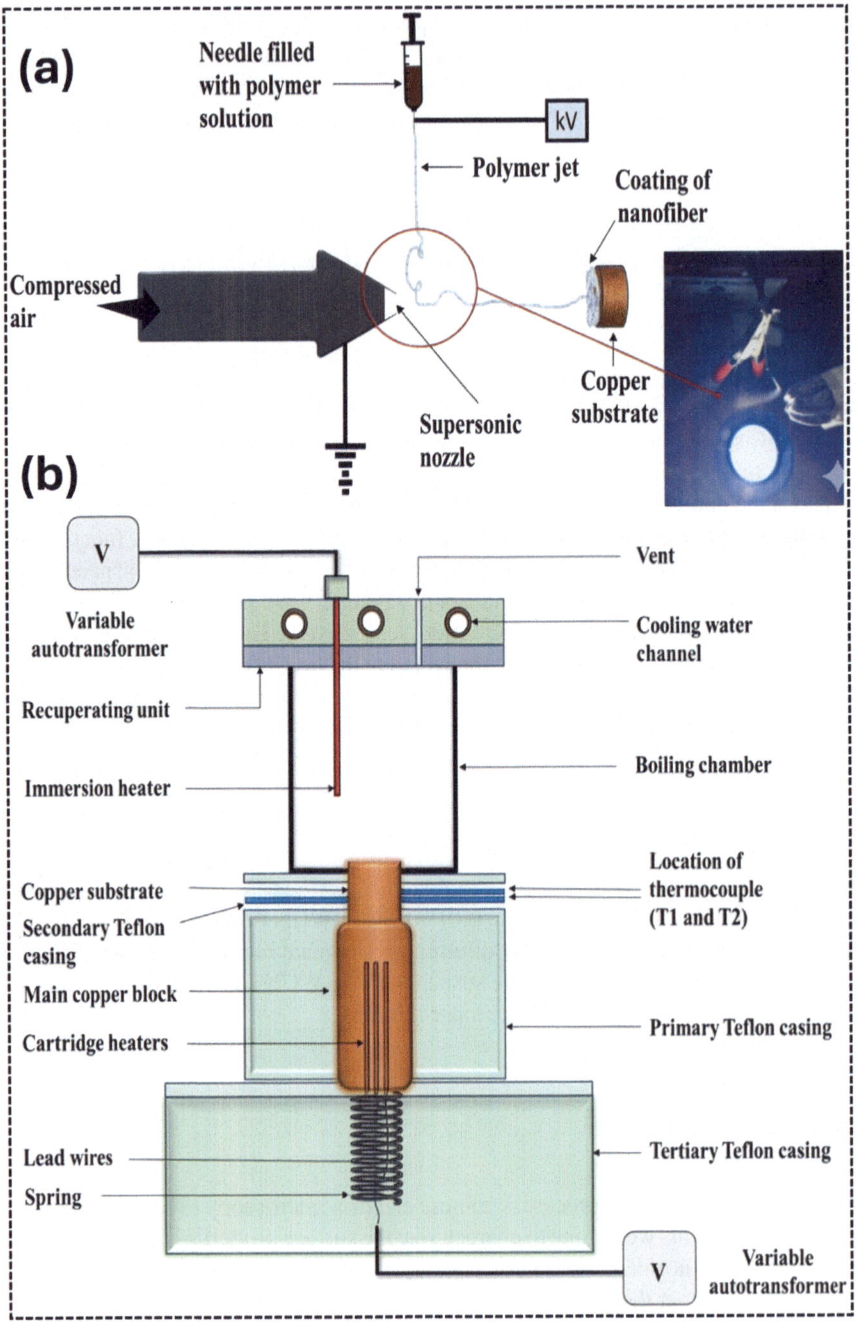

Fig. 1 Schematic representation of **a** nanofiber coating on substrate via supersonic solution blowing (SSB) and **b** pool boiling experimental setup

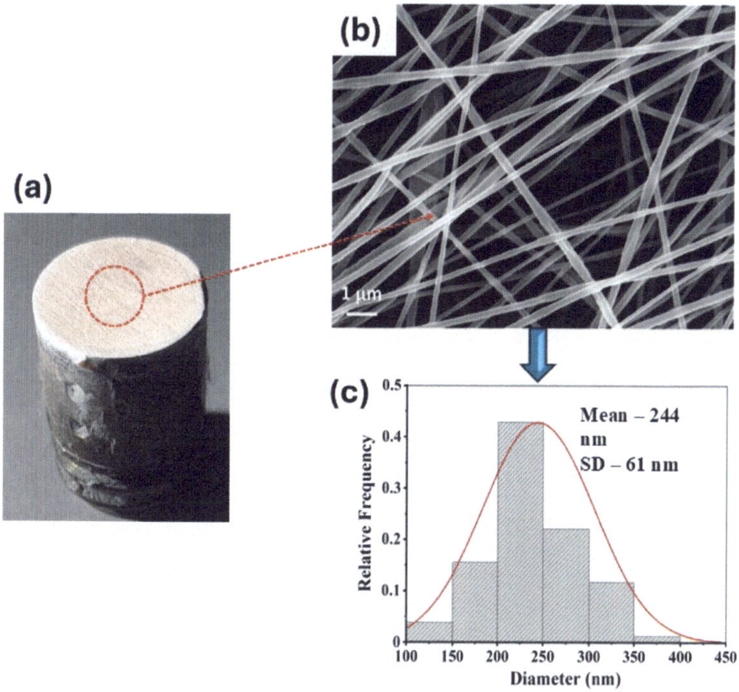

Fig. 2 **a** Represents the coated image of substrate **b** SEM image of nanofibers and **c** diameter distribution of nanofibers

137.7°. Figure 2a demonstrates the image of coated substrate, (b) SEM image of PVDF nanofibers and (c) represents the diameter distribution of the nanofibers.

3.5 Nanofluid Property

Nanofluids' characteristics depend on the properties of the base fluid and the nanoparticles. To estimate the fluid properties, the mass fraction was converted into the volumetric fraction using the correlations [33]:

$$\varphi_v = 1 \Big/ \left(\frac{1-\varphi_m}{\varphi_m}\right)\frac{\rho_p}{\rho_f} + 1 \tag{1}$$

The density of the nanofluid [28] can be estimated by using:

$$\rho_{nf} = \rho_f(1-\varphi_v) + \rho_p\varphi_v \tag{2}$$

Using the above relation, specific heat [28] can be estimated by:

Table 1 Thermophysical properties of working fluids (DI water [36], SWCNT [37, 38])

Properties/Fluid	DI Water	SWCNT	NF
ρ (kg/m^3)	958.3	1900	958.34
C_p (kJ/kg K)	4.22	0.6	4.22
κ (W/m K)	0.67	3500	0.68
μ (mPa s)	0.27	–	0.27

$$\left(\rho c_p\right)_{nf} = \rho_f c_{p,f}(1 - \varphi_v) + \rho_p c_{p,p}\varphi_v \tag{3}$$

The thermal conductivity of the nanofluid can be estimated using the Hamilton-Crosser model [34]:

$$k_{nf} = k_f \frac{k_p + (n-1)k_f + (n-1)\varphi_v\left(k_p - k_f\right)}{k_p + (n-1)k_f - \varphi_v\left(k_p - k_f\right)} \tag{4}$$

If ($\varphi_v \leq 2\%$), viscosity of the nanofluid can be determined by using the relation below [35]:

$$\mu_{nf} = (1 + 2.5\varphi_v)\mu_f \tag{5}$$

By using the above relations, the thermophysical properties of the working fluids are calculated. Refer to Table 1.

3.6 Experimental Setup

Figure 1b illustrates the schematic of the pool boiling experimental apparatus. The key components include: (a) Three Teflon casings, primary Teflon casing, secondary Teflon casing, and tertiary Teflon casing, to prevent thermal losses (b) main copper cylindrical block (c) the copper substrate affixed to the copper cylindrical block (d) three cartridge heaters are inserted into the main copper cylindrical block to provide uniform heat (e) boiling chamber constructed from the aluminum block and glass window [for observational purpose] (f) vapor condensing unit with an immersion heater within. The copper substrate has a dimension of 12.3 mm diameter and 19 mm length. Two holes, each 1.5 mm in diameter, were drilled into the center of the copper substrate, 3.8 mm apart, to allow for the insertion of thermocouples to measure temperatures (T1 and T2). To supply the uniform heat to the copper block, three cartridge heaters (100 W each) were vertically inserted into the copper block through its base. Primary and secondary teflon blocks were employed to minimize thermal losses from both i.e., the copper substrate and the copper block. A cast iron spring was positioned within a tertiary teflon casing, and the primary copper cylinder and the inserted cartridge heater assembly were attached above it. The heaters were fitted into the cylindrical copper block, and the spring was pressed against its bottom to ensure reduced heat loss due to appropriate contact. A thermally conductive paste (Arctic Silver™) was applied at the interface between the copper cylinder and the copper substrate to minimize thermal contact resistance. The boiling chamber measuring $40 \times 60 \times 76$ mm^3 was constructed from aluminum, with glass installed on its two vertical faces for visibility. An O-ring was utilized to bridge the space between the copper substrate and the secondary teflon casing. Above the boiling chamber, a vapor condensing apparatus was constructed from aluminum, and several copper tubes were put through it, allowing cold water to run continually. To prevent pressure buildup in the chamber, a small 1 mm (diameter) hole was machined at the corner region of the recuperating assembly to facilitate the vapor release. The working fluid was maintained at saturation temperature by placing a 100 W immersion heater into the boiling chamber through the top of the recuperating unit.

3.7 Pool Boiling Experiment

The copper substrate was smoothed sequentially using 1000, 1200, and 2500 grit sandpaper. After assembling the pool boiling set-up, 100 g of DI water or nanofluid was poured into the boiling chamber. To collect temperatures at two distinct sites, two K-type thermocouples (T_1 and T_2, Omega Engineering) were embedded within the copper substrate and coupled via a thermometer (see Fig. 1b for locations). Electric power supply to the immersion heater at the top and cartridge heater assembly were provided by two distinct variable autotransformers. Throughout the trials, the immersion heater's voltage was consistently kept at 120 V. Initially, a heater assembly connected to variable autotransformers was operated at 40 V, and then after some time it was increased by 10 V till the heater assembly reaches close to saturation temperature of DI Water (100 °C). During this duration immersion heater degassed the base fluid (DI Water). After the degassing period, voltage was increased by 5 V and was kept at this voltage for 10 min, so that the surface temperature reached steady state conditions. This cycle was repeated till the 140 V. At 140 V we have observed the different heat flux for different surface-fluid combinations. At that point the experiment was stopped.

4 Results and Discussion

During the experiment, a one-dimensional heat conduction model was assumed along the axial direction of the copper substrate. The surface temperature (T_s) was calculated by assuming the linear temperature distribution along the length of the copper substrate [13]:

$$T_s = T_2 - \left(\frac{T_1 - T_2}{\Delta x_{12}}\right)\Delta x_{2s} \tag{6}$$

where Δx_{12} and Δx_{12} are the distances between T_1 and T_2 and T_s and surface respectively. The heat flux supplied to the surface is estimated by using:

$$q'' = k_{cu}\left(\frac{T_1 - T_2}{\Delta x_{12}}\right) \tag{7}$$

The surface superheat was estimated by using, $\Delta T = T_s - T_{sat}$. The BHTC (h) was calculated by $h = q'' / \Delta T$.

4.1 Pool Boiling of DI Water and Nanofluid on Bare Surface

Figure 3 shows the average experimental data of pool boiling of DI water and nanofluid (NF) on the bare surface. The data clearly suggests an enhancement in the average heat flux and BHTC for NF as compared to DI Water. The onset of nucleate boiling (ONB) was observed at 9.07 and 3.20 °C for DI water and NF respectively. As the degassing and subsequent boiling took place, some SWCNT is possibly deposited over the copper surface. This nanoparticle deposition could form a smooth microlayer at the heated surface, reducing the radius of surface defects and shifting the ONB to a lower surface superheat than DI water. For bare/NF combination, the maximum heat flux in the nucleate boiling region was obtained at 949.73 kW/m^2 at 11.67 °C surface superheat, while for bare/DI water combination, it was observed at 517.07 kW/m^2 at 21.65 °C surface superheat. This indicates 83.67% enhancement in bare/NF combination as compared to bare/DI water combination. To predict the boiling behavior and the effect of surface on the micro-scale fluid flow governed by the nucleate boiling, Rohsenow's correlation [39] is used, where the terms have their usual meaning and is heavily influenced by the surface characteristics:

$$q = \mu_l h_{fg}\left(c_{p,l}\Delta T \Big/ h_{fg} C_{sf} \overset{n}{\mathrm{Pr}}\right)^m \left[g(\rho_l - \rho_v)/\sigma\right]^{0.5} \tag{8}$$

In above equation Prandtl number is calculated using $\mathrm{Pr} = \mu_l c_{p,l} / k_l$ and m and n are fitting parameters (according to Rohsenow's correlation m = 3 and n = 1.7), C_{sf} is the numerical constant which depends on the combined effect of the surface and working fluid. Satyam et al. [33] reported the value of C_{sf} and n for ethanol/copper combination as 0.00185 and 1.45, respectively. In the present work the value of C_{sf} for bare/DI water and bare/NF combination was 0.0185 and 0.0075, respectively. The value 'n', a parameter that dictates the effect of Prandtl on the overall growth of the boundary layer, was calculated to be 0.82 and 1.11 for bare/DI water and bare/NF combination, respectively.

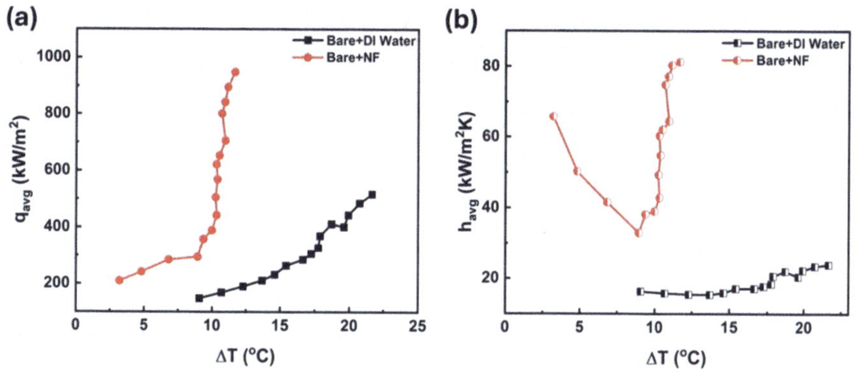

Fig. 3 Pool boiling behavior of DI Water and nanofluid (NF, as abbreviated in the image) on bare surface—**a** average heat flux vs surface superheat and **b** average BHTC vs surface superheat

4.2 *Pool Boiling of DI Water and Nanofluid on Nanotextured Surface*

Figure 4 shows the experimental data of pool boiling of DI water and NF on the nanotextured (NT) surface. The ONB for DI water/NT surface and NF/NT surface was obtained at 3.23 and 4.48 °C, respectively. According to the published research [26], the reason behind shifting of ONB is deposition of nanoparticles during micro-layer evaporation at the heated surface. The BHTC for NT/DI water surface and NT/NF surface was found to be 121.60 kW/m^{2}K and 79.35 kW/m^{2}K respectively, indicating a ~53.24% enhancement in case of NT/DI water as compared to NT/NF, respectively. The previously mentioned fitting parameters (C_{sf}, n) for NT/DI Water surface and NT/NF surface were found to be 0.0066, 0.80 and 0.01, 1.0099 respectively. As evident from Fig. 4, the nucleate boiling exhibits the higher maximum heat flux values for NT/DI water as compared to NT/NF. The maximum heat flux for NT/DI water and NT/NF surface combinations were found to be 1255.76 kW/m^2 and 1065.81 kW/m^2, respectively.

Visibly for both water and nanofluid, the nanotextured surface performed better than pure copper surface, despite the polymer fibrous network being insulative in nature. However, it is to be noted that due to individual scale of nanofibers and the layer's overall thickness being small, the transient time helps to attain similar temperature as the surface is much smaller compared to the steady/process time. This ensured that growing embryos experienced temperature similar to the surface as they grew from the fibrous network. This overheating of the bubble-surrounding fluid didn't favor bubble growth and/or coalescence. This was aided by the high contact angle of the fibrous network. Poor wettability favored vapor bubble escape from the surface due to low surface energy Additionally it had been discussed in literature [12] that vapor bubbles tend to get caged between the fibrous network as they grow from the surface through the porous architecture. The elastic nature of the liquid interface

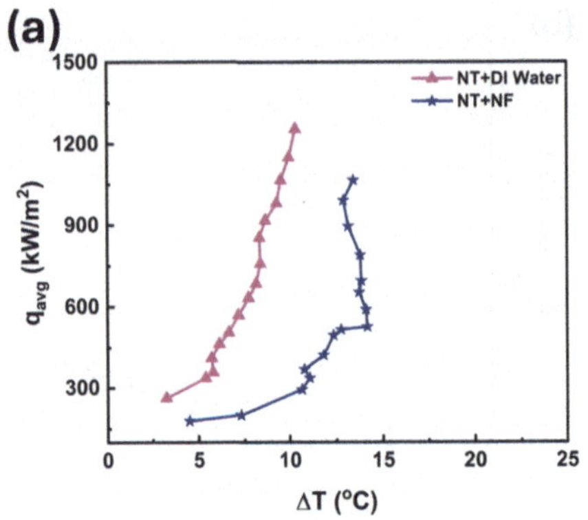

Fig. 4 Pool boiling behavior of DI Water and nanofluid (NF, as abbreviated in the image) on nanotextured (NT, as abbreviated in the image) surface—**a** average heat flux vs surface superheat and **b** average BHTC vs surface superheat

of the vapor bubble allows bubble chopping. This ensured smaller vapor bubbles releasing from the surface at higher frequency and significant heat removal at the nucleation stage. This was evident in both Figs. 3a and 4a where a significant surface superheat reduction can be observed. However, for nanofluids, fouling effects might reduce the rapid nucleation behavior, mostly due to enhanced wettability, stemming from the presence of CNTs and due to reduction of available pores. Since, this study focused only on nucleation behavior up to a certain power supply, whether the boiling behavior of nanofluid combinations would show an overall better boiling behavior, in terms of both nucleation and critical heat flux, remains a question to explore.

5 Conclusion

In this study, pool boiling phenomena of DI water and SWCNT-based nanofluid was observed on bare copper and nanofibre textured surface. The NT surface was shown to be more efficient in promoting nucleate boiling than the bare surface. The combination of NT/DI water exhibited 143% enhancement in the heat flux in the nucleate boiling region compared to bare/DI water combination. Moreover, it also enhances the BHTC by nearly 5.1 times as compared to bare/DI water combination. The presence of SWCNTs in NF didn't affect the thermophysical properties of DI water, however, it also improved the boiling behavior in the bare surface significantly, with nearly 106% improvement in heat flux removal with nanotexture combination. The fouling effects of CNT could better the boiling behavior by enhancing overheating of the fluid in the vicinity of the surface. However for nanotexturing, the fouling effects could encase the pores, reducing the overall cages to limit nucleation boiling.

Acknowledgements R.P acknowledges the scholarship received from Ministry of Education (MoE), Govt of India. R.P. and S.S.R gratefully acknowledge the Central Research Facility of IIT Delhi and the Department of Material science, IIT Delhi, for permitting to use the sophisticated instrument facility.

References

1. Bar-Cohen A, Iyengar M (2002) Design and optimization of air-cooled heat sinks for sustainable development. IEEE Trans Compon Packag Technol 25(4):584–591
2. Choi J, Jeong M (2016) Compact, lightweight, and highly efficient circular heat sink design for high-end PCs. Appl Therm Eng 92:162–171
3. Faghri A (1995) Heat pipe science and technology. Taylor & Francis
4. Dhir VK (1998) Boiling heat transfer. Annu Rev Fluid Mech 30(1):365–401
5. Kim J (2009) Review of nucleate pool boiling bubble heat transfer mechanisms. Int J Multiph Flow 35(12):1067–1076
6. Kim DE, Yu DI, Jerng DW, Kim MH, Ahn HS (2015) Review of boiling heat transfer enhancement on micro/nanostructured surfaces. Exp Thermal Fluid Sci 66:173–196
7. Betz AR, Jenkins J, Attinger D (2013) Boiling heat transfer on superhydrophilic, superhydrophobic, and superbiphilic surfaces. Int J Heat Mass Transf 57(2):733–741
8. Zupančič M, Steinbücher M, Gregorčič P, Golobič I (2015) Enhanced pool-boiling heat transfer on laser-made hydrophobic/superhydrophilic polydimethylsiloxane-silica patterned surfaces. Appl Therm Eng 91:288–297
9. Jo H, Ahn HS, Kang S, Kim MH (2011) A study of nucleate boiling heat transfer on hydrophilic, hydrophobic and heterogeneous wetting surfaces. Int J Heat Mass Transf 54(25–26):5643–5652
10. Jun S, Sinha-Ray S, Yarin AL (2013) Pool boiling on nano-textured surfaces. Int J Heat Mass Transf 62(1):99–111. https://doi.org/10.1016/J.IJHEATMASSTRANSFER.2013.02.046
11. Sahu RP, Sinha-Ray S, Sinha-Ray S, Yarin AL (2015) Pool boiling on nano-textured surfaces comprised of electrically-assisted supersonically solution-blown, copper-plated nanofibers: Experiments and theory. Int J Heat Mass Transf 87. https://doi.org/10.1016/j.ijheatmasstransfer.2015.04.009
12. Sahu RP, Sinha-Ray S, Sinha-Ray S, Yarin AL (2016) Pool boiling of Novec 7300 and self-rewetting fluids on electrically-assisted supersonically solution-blown, copper-plated nanofibers. Int J Heat Mass Transf 95. https://doi.org/10.1016/j.ijheatmasstransfer.2015.11.094
13. Sinha-Ray S, Zhang W, Sahu RP, Sinha-Ray S, Yarin AL (2017) Pool boiling of Novec 7300 and DI water on nano-textured heater covered with supersonically-blown or electrospun polymer nanofibers. Int J Heat Mass Transf 106:482–490. https://doi.org/10.1016/j.ijheatmasstransfer.2016.08.101
14. Jo HS et al (2017) Supersonically sprayed, triangular copper lines for pool boiling enhancement. Int J Heat Mass Transf 113:210–216
15. An S et al (2016) Supersonically sprayed reduced graphene oxide film to enhance critical heat flux in pool boiling. Int J Heat Mass Transf 98:124–130
16. Ciloglu D, Bolukbasi A (2015) A comprehensive review on pool boiling of nanofluids. Appl Therm Eng 84:45–63
17. Fang X, Chen Y, Zhang H, Chen W, Dong A, Wang R (2016) Heat transfer and critical heat flux of nanofluid boiling: a comprehensive review. Renew Sustain Energy Rev 62:924–940
18. Zimmermann M, Heinz M, Sielaff A, Gambaryan-Roisman T, Stephan P (2020) Influence of system pressure on pool boiling regimes on a microstructured surface compared to a smooth surface. Exp Heat Transf 33(4):318–334
19. Pioro IL, Rohsenow W, Doerffer SS (2004) Nucleate pool-boiling heat transfer. I: review of parametric effects of boiling surface. Int J Heat Mass Transf 47(23):5033–5044

20. Bourdon B, Di Marco P, Riobóo R, Marengo M, De Coninck J (2013) Enhancing the onset of pool boiling by wettability modification on nanometrically smooth surfaces. Int Commun Heat Mass Transf 45:11–15
21. Moghadasi H, Saffari H, Malekian N (2021) Experimental and semi-analytical investigation of heat transfer in nucleate pool boiling by considering surface structuring methods. Exp Heat Transf 34(4):293–313
22. Nirgude VV, Sahu SK (2019) Enhancement in nucleate pool boiling heat transfer on nano-second laser processed copper surfaces. Exp Heat Transf 32(6):566–583
23. Weickgenannt CM et al (2011) Inverse-Leidenfrost phenomenon on nanofiber mats on hot surfaces. Phys Rev E Stat Nonlinear Soft Matter Phys 84(3). https://doi.org/10.1103/PhysRevE.84.036310
24. Gupta SK, Misra RD (2019) An experimental investigation on pool boiling heat transfer enhancement using Cu–Al_2O_3 nano-composite coating. Exp Heat Transf 32(2):133–158
25. Kim H, Kim J, Kim MH (2006) Effect of nanoparticles on CHF enhancement in pool boiling of nano-fluids. Int J Heat Mass Transf 49(25–26):5070–5074
26. Das SK, Putra N, Roetzel W (2003) Pool boiling characteristics of nano-fluids. Int J Heat Mass Transf 46(5):851–862
27. Hegde RN, Rao SS, Reddy RP (2012) Investigations on boiling-induced nanoparticle coating, transient characteristics, and effect of pressure in pool boiling heat transfer on a cylindrical surface. Exp Heat Transf 25(4):323–340
28. Bang IC, Chang SH (2005) Boiling heat transfer performance and phenomena of Al_2O_3–water nano-fluids from a plain surface in a pool. Int J Heat Mass Transf 48(12):2407–2419
29. Singh SK, Sharma D (2022) Pool boiling heat transfer enhancement using nanoparticle coating on copper substrate. In: Handbook of green and sustainable nanotechnology: fundamentals, developments and applications. Springer, pp 1–23
30. Dahariya S, Patel N, Egbo MK, Hwang G, Betz AR (2020) High-pressure pool-boiling heat transfer enhancement mechanism on sintered-particle wick surface. Front Mech Eng 5:71
31. Yao Z, Lu Y-W, Kandlikar SG (2012) Pool boiling heat transfer enhancement through nanostructures on silicon microchannels. J Nanotechnol Eng Med 3(3):31002
32. Gajghate SS, Barathula S, Das S, Saha BB, Bhaumik S (2020) Experimental investigation and optimization of pool boiling heat transfer enhancement over graphene-coated copper surface. J Therm Anal Calorim 140(3):1393–1411
33. Thakur SS, Chandel SS, Kakoria A, Sinha-Ray S (2022) Enhancement in pool boiling heat transfer of ethanol and nanofluid on novel supersonic nanoblown nanofiber textured surface. Experimental Heat Transfer 35(4):516–532. https://doi.org/10.1080/08916152.2021.1919243
34. Yang L, Xu X (2017) A renovated hamilton-crosser model for the effective thermal conductivity of CNTs nanofluids. Int Commun Heat Mass Transfer 81:42–50
35. Bashirnezhad K et al (2016) Viscosity of nanofluids: a review of recent experimental studies. Int Commun Heat Mass Transfer 73:114–123
36. Carey VP (2020) Liquid-vapor phase-change phenomena: an introduction to the thermophysics of vaporization and condensation processes in heat transfer equipment. CRC Press
37. Muhammad S, Ali G, Shah Z, Islam S, Hussain SA (2018) The rotating flow of magneto hydrodynamic carbon nanotubes over a stretching sheet with the impact of non-linear thermal radiation and heat generation/absorption. Appl Sci 8(4):482
38. Said Z, Saidur R, Sabiha MA, Rahim NA, Anisur MR (2015) Thermophysical properties of single wall carbon nanotubes and its effect on exergy efficiency of a flat plate solar collector. Sol Energy 115:757–769. https://doi.org/10.1016/j.solener.2015.02.037
39. Rohsenow WM (1952) A method of correlating heat-transfer data for surface boiling of liquids. Trans Am Soc Mech Eng 74(6):969–975

Experimental Investigation of Electric Discharge Machining of Nimonic C-263 Using Circular Shaped Copper Tool Electrode

Sujoy Chakraborty, Uttam Kumar Mandal, Debashis Podder, Lokesh Kumar Ranjan, and Divice Boro

Abstract This study focuses on Electric Discharge Machining (EDM) of Nimonic C-263, a nickel-based superalloy known for its high strength, corrosion resistance, and performance at elevated temperatures. The research emphasizes the use of a circular copper electrode to evaluate machining performance indicators such as Material Removal Rate (MRR) and Tool Wear Rate (TWR). Key EDM process variables—including discharge current, pulse on-time, and pulse off-time—were systematically varied to observe their impact. Results indicate that discharge current is the most influential parameter affecting MRR, followed by pulse on-time and pulse off-time. A Taguchi L9 orthogonal array was utilized to optimize process conditions. The best machining setup was determined to be a 10 A current, 7 μs pulse on-time, and 5 μs pulse off-time. These results offer practical guidance for efficient EDM of Nimonic C-263 and contribute to a deeper understanding of the machining mechanisms involved.

Keywords EDM · MRR · TWR · Taguchi and orthogonal array

1 Introduction

Nimonic C-263 is a superior nickel-based superalloy valued for its remarkable mechanical strength, resistance to corrosion, and thermal stability at high temperatures. Due to these characteristics, it is extensively used in critical sectors such as aerospace and power generation, where materials are exposed to harsh and extreme operating conditions. However, these same properties also make the material difficult to machine using traditional manufacturing techniques. To overcome these machining challenges, advanced non-conventional techniques such as Electric Discharge Machining (EDM) are often employed. EDM is a thermoelectric

S. Chakraborty (✉) · U. K. Mandal · D. Podder · L. K. Ranjan · D. Boro
Production Engineering Department, NIT Agartala, Agartala, India
e-mail: ersujoymech05@gmail.com

A. Biswas et al. (eds.), *Recent Advancements in Artificial Intelligence and Optimization*, Lecture Notes in Mechanical Engineering,
https://doi.org/10.1007/978-981-95-9268-5_8

process that removes the material from the workpiece through controlled electrical discharges. Unlike traditional machining methods that rely on physical cutting forces, EDM is ideal for processing hard, brittle, or thermally resistant materials, as well as for fabricating intricate geometries. In this process, repeated electrical sparks are generated between a tool and the workpiece, both immersed in a dielectric fluid. These sparks produce intense localized heat, causing the material to melt and evaporate at the spark location. This research explores the application of EDM for machining Nimonic C-263, with a specific focus on performance metrics like Material Removal Rate (MRR) and Tool Wear Rate (TWR). The study utilizes a circular copper electrode, and investigates the impact of varying electrical parameters—namely discharge current, pulse on-time, and pulse off-time—on machining outcomes. The primary aim of this work is to establish optimal EDM settings for effective machining of Nimonic C-263 and to improve understanding of the fundamental mechanisms involved. A Taguchi L9 orthogonal array was adopted for experimental design and parameter optimization. Results reveal that discharge current plays the most crucial role in determining MRR, followed by pulse on-time and pulse off-time. Experimental analysis led to the identification of optimal conditions: a discharge current of 10 A, a pulse on-time of 7 μs, and a pulse off-time of 5 μs. These parameter settings offer a balanced trade-off between maximizing MRR and minimizing TWR, thus enhancing the overall machining performance. The study provides valuable insights into the capabilities and constraints of EDM when applied to superalloys like Nimonic C-263. Additionally, the findings deliver practical recommendations for industrial applications involving this alloy, contributing to advancements in precision machining of advanced materials.

2 Literature Review

Research on Electric Discharge Machining (EDM) of superalloys has shown that machining outcomes are significantly influenced by factors such as electrode material, tool geometry, and process variables. Copper is widely used as an electrode due to its high thermal and electrical conductivity, which aids in efficient spark discharge and material removal. Electrodes with a circular profile are particularly beneficial for uniform energy dispersion and minimizing tool wear irregularities. [1] explored the machinability of Nimonic C-263 using wire EDM and assessed how various process parameters influenced aspects like surface finish, spark gap, and wire wear. Their approach, which combined mathematical modeling and a multicut strategy, led to the identification of efficient machining conditions. Amitava and Pramanik [2] evaluated the surface characteristics of Nimonic C-263 post-WEDM, focusing on roughness, hardness, and microstructural changes. They emphasized the need for secondary processes like grinding and etching to improve the final surface quality by removing recast layers. In another contribution, [3] investigated the use of Powder-Mixed EDM (PMEDM), integrating titanium powder in a deionized water dielectric, and applied grey relational analysis (GRA) for optimization. Their results confirmed notable improvements in MRR and surface finish due to the powder additives. References [4–6, 10] expanded on EDM optimization by employing artificial intelligence techniques such as artificial neural networks (ANN), as well as Taguchi-based and hybrid approaches to enhance surface quality and MRR. Environmentally focused research by Dewan and Kundu [9, 15] introduced green dielectric fluids for EDM and utilized multi-criteria decision-making methods to optimize process conditions, highlighting discharge current as a key influencing factor. Recent advancements include a servo-controlled levitation system for micro-EDM proposed by [11] to enhance accuracy and stability. [12] conducted a comparative machinability analysis of different nickel-based alloys using WEDM. Additional studies by [13, 14, 17] have contributed to the optimization of EDM and WEDM processes through statistical modeling, software-assisted tuning, and sustainability-driven techniques. Taken together, the reviewed literature underscores the value of precisely adjusting EDM parameters to optimize the machining of challenging materials like Nimonic C-263. Despite substantial progress, there remains a need for further research focused specifically on this alloy. The present work aims to build on these insights by delivering an experimentally validated EDM strategy tailored to the high-precision requirements of Nimonic C-263.

3 Experimental Setup

The experimental work was carried out using a Sparkonix F25 die-sinking EDM machine, as illustrated in Fig. 1a. The system comprises key components such as a servo head, dielectric fluid reservoir, circulation pump, and control unit. The servo mechanism is responsible for precisely maintaining the gap between the electrode and the workpiece to ensure stable sparking conditions. Nimonic C-263 samples, each measuring 20 mm × 20 mm × 10 mm, were used as shown in Fig. 1b. A 10 mm diameter copper electrode with a circular profile (Fig. 1c) served as the tool. The main process variables adjusted during experimentation included discharge current (I), pulse on-time (T_{on}), and pulse off-time (T_{off}). Parameter levels were selected based on prior research and trial runs.

4 Methodology

The study employed the Taguchi design methodology to plan and analyze the experiments, making use of an L9 orthogonal array suitable for testing combinations of multiple factors at different levels. This statistical approach allowed the efficient examination of the influence of EDM parameters on responses such as Material Removal Rate (MRR) and Tool Wear Rate (TWR). MRR was computed by measuring the weight difference of the workpiece before and after machining, whereas TWR was obtained by evaluating the reduction in mass of the tool electrode across each

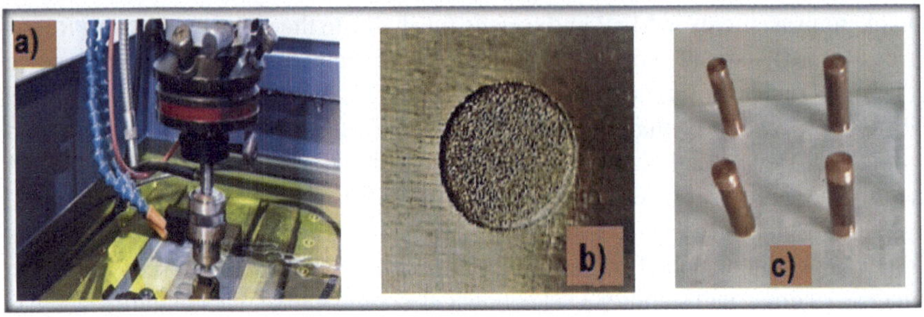

Fig. 1 **a** Experimental set up; **b** EDMed workpiece sample; **c** Circular shaped electrode

Table 1 Level of process parameters

Process parameter	Pulse on-time (Ton) (μs)	Pulse off-time (Toff) (μs)	Peak current (Ip) (A)
Level 1	5–100	5–30	8
Level 2	6–150	6–40	9
Level 3	7–200	7–75	10

Table 2 L-9 Orthogonal array and experimental results for MRR and TWR

Exp no.	T_{on} (μs)	T_{off} (μs)	I_p (A)	MRR (mg/min)	TWR (mg/min)
1	100	30	8	0.0067	0.00074
2	100	40	9	0.0069	0.00075
3	100	75	10	0.007	0.00077
4	150	30	9	0.0072	0.00078
5	150	40	10	0.0073	0.0008
6	150	75	8	0.0068	0.00078
7	200	30	10	0.0076	0.00085
8	200	40	8	0.0071	0.00082
9	200	75	9	0.0069	0.00083

trial. The process variables and their respective levels used for the experimental design are detailed in Table 1.

5 Results and Discussion

Nine experimental trials were conducted as part of this investigation. The Material Removal Rate (MRR) and Tool Wear Rate (TWR) were calculated based on mass changes recorded using a precision weighing balance. The differences in weight before and after machining were used to quantify material removal and tool wear over the total machining time. The formulas used to determine machining responses are as follows:

$$\textbf{MRR}(\textbf{mg/min}) = (\text{Final weight of workpiece} - \text{Initial weight})/\text{Total machining time} \quad (1)$$

$$\textbf{TWR}(\textbf{mg/min}) = (\text{Initialweightoftool} - \text{Finalweight})/\text{Totalmachiningtime} \quad (2)$$

Each experiment was repeated three times, and the average values were used for analysis. Signal-to-noise (S/N) ratios were computed to evaluate the robustness of process parameters under variable conditions (Table 2).

The influence of each process parameter on MRR and TWR is analyzed in the following subsections.

5.1 Analysis of Material Removal Rate (MRR)

MRR exhibited an upward trend with increasing discharge current and longer pulse on-times, as both parameters contribute to higher energy discharge during machining. This enhanced thermal energy promotes greater material melting and vaporization, thereby raising the removal rate. However, excessively high settings may introduce instability, leading to a drop in MRR.

Figure 2 presents the influence of discharge current on the Material Removal Rate (MRR), indicating a steady rise in MRR as the current increases from 8 to 10 A, with a maximum value of 0.0076 mg/min observed at 10 A. This improvement is linked to elevated spark energy and breakdown voltage, which facilitate enhanced melting and evaporation of the workpiece surface. In contrast, a gradual increase in pulse-off time from 5 to 7 μs results in a noticeable decline in MRR. Figure 3 displays the corresponding signal-to-noise (S/N) ratio for MRR. Since pulse-off time represents the interval without spark generation, extending it prolongs the overall machining cycle and lowers spark frequency, thus reducing material removal. Meanwhile, increasing the pulse-on time from 5 to 7 μs contributes positively to MRR by supplying greater energy per discharge.

5.2 Analysis of Tool Wear Rate (TWR)

TWR followed a trend similar to MRR, where higher values of discharge current and pulse on-time resulted in elevated tool wear. This is primarily due to increased thermal loading on the electrode. Nonetheless, certain combinations allowed acceptable tool

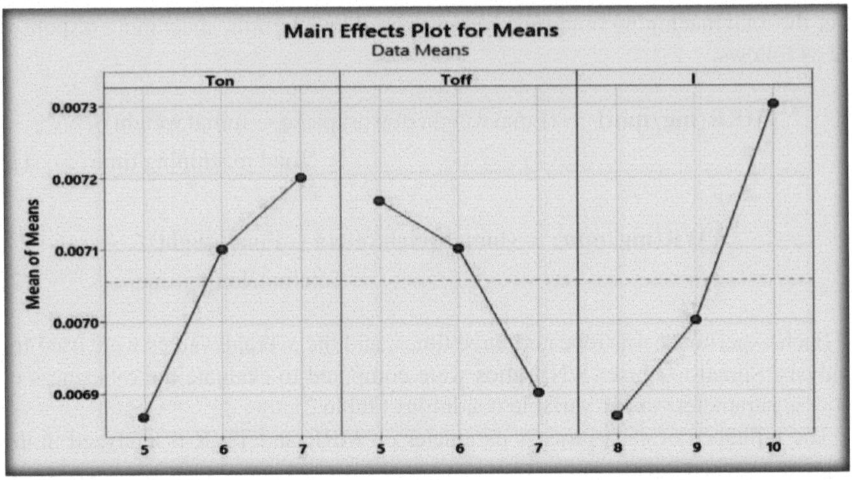

Fig. 2 MRR mean effect plot

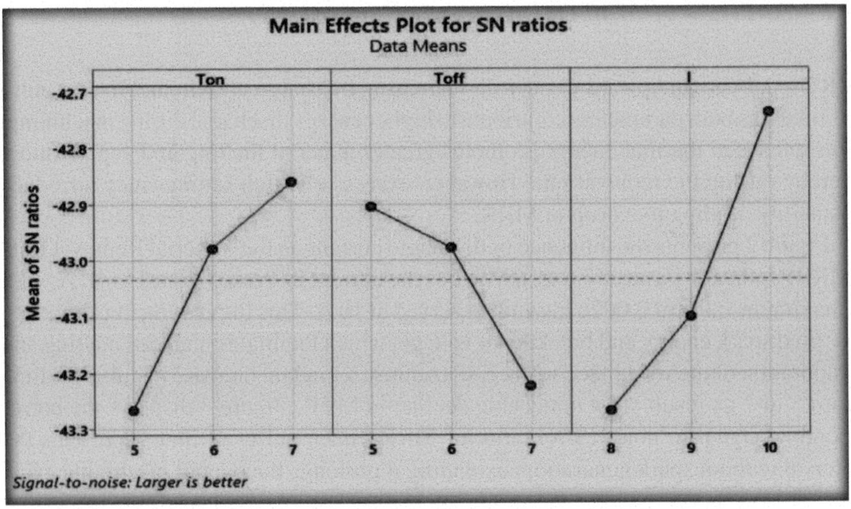

Fig. 3 MRR SN ratio plot

wear while maintaining reasonable MRR levels, enabling a practical compromise between efficiency and tool life.

Figure 4 demonstrates that, wear occurs when the electrode is degraded by the thermoelectric heating technique used during the EDM process. As the off time increases, the spark has less time to make contact with the work, resulting in a rise in TWR. The Tool Wear Rate (TWR) changes very little at the greatest T_{off} values, perhaps as a result of less debris adhering to the tool surface that is further assisted by rotation of the tool. Signal to Noise ratio for TWR is shown in Fig. 5. The increase in TWR is primarily associated with higher discharge currents, which facilitate the formation of plasma channels and raise the tool surface temperature due to melting and thermal erosion.

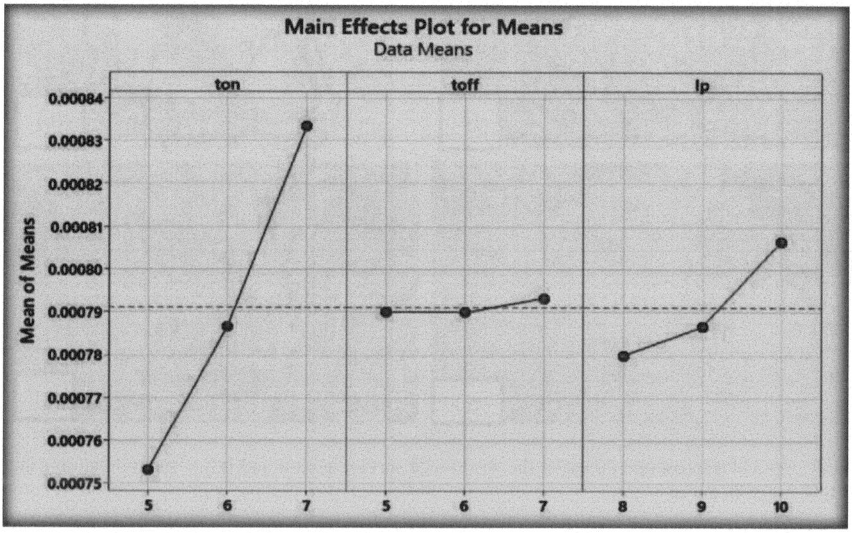

Fig. 4 TWR mean effect plot

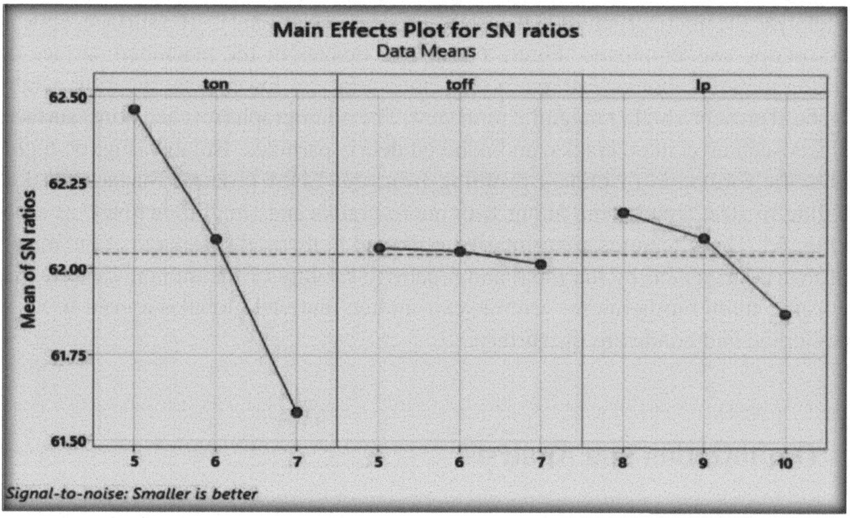

Fig. 5 TWR SN ratio plot

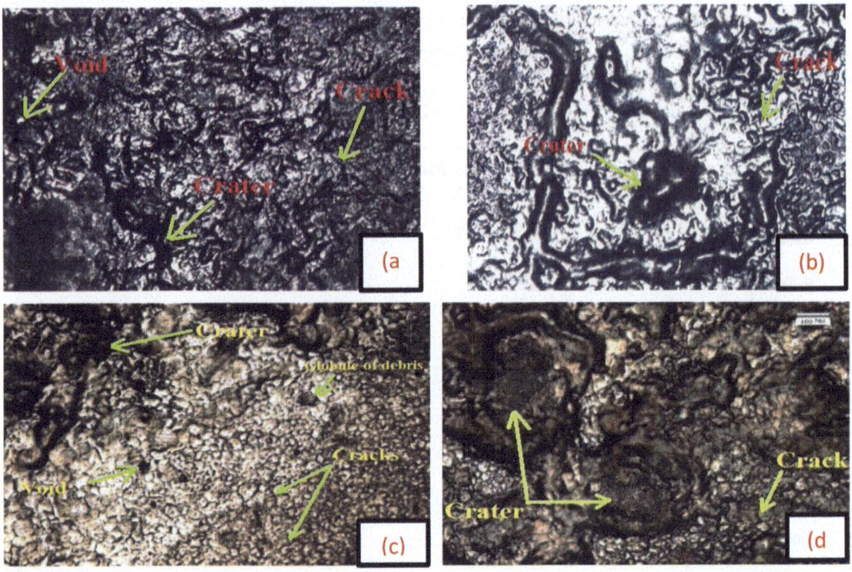

Fig. 6 Optical microscope image of the workpiece surface with **a** and **b** 5 × magnification **c** and **d** 10 × magnification

5.3 Microstructure Analysis

To analyze the microstructure of the EDM-processed Nimonic C-263 surface, optical microscopy was employed. Figure 6 displays images of the machined surface at varying magnifications—Fig. 6a, b show the surface at lower magnification, while Fig. 6c, d present a higher magnification view. The micrographs reveal distinct surface defects such as craters, cracks, and adhered debris particles. Notably, Fig. 6c highlights the formation of craters resulting from the EDM process. Surface cracks, ranging in size, are evident, along with micro-cracks and small blowholes, as seen in Fig. 6d. These features are primarily attributed to thermal shrinkage of the material matrix, triggered by the rapid and repetitive cooling of the molten zone by the dielectric fluid. Furthermore, several semi-molten material globules appear to have resolidified and bonded to the surface.

6 Optimization and Analysis

Taguchi analysis was conducted on the experimental data to identify the optimal combination of process parameters. Both the mean values and the signal-to-noise (S/N) ratios were computed using MINITAB software. The influence of each factor on Material Removal Rate (MRR) is summarized in Table 3 for mean values and Table 4

Table 3 Response table for means (MRR)

Level	T_{on} (μs)	T_{off} (μs)	I (A)
1	0.006867	0.007167	0.006867
2	0.0071	0.0071	0.007
3	0.0072	0.0069	0.0073
Delta	0.000333	0.000267	0.000433
Rank	2	3	1

Table 4 Response table for S/N ratio for MRR

Level	T_{on} (μs)	T_{off} (μs)	I (A)
1	–43.27	–42.91	–43.27
2	–42.98	–42.98	–43.1
3	–42.86	–43.22	–42.74
Delta	0.41	0.32	0.53
Rank	2	3	1

for S/N ratios. Among the parameters studied, peak current had the most pronounced impact on MRR, followed by pulse on-time (T_{on}) and pulse off-time (T_{off}). Similarly, for Tool Wear Rate (TWR), the mean response is detailed in Table 5 and the S/N analysis in Table 6. In this case, pulse on-time (T_{on}) was the most dominant factor, with peak current and pulse off-time contributing to a lesser extent.

Based on the Taguchi optimization approach, the ideal EDM parameters for machining Nimonic C-263 were identified as a discharge current of 10 A, a pulse on-time of 7 μs, and a pulse off-time of 5 μs. This combination provided an effective

Table 5 Response table for means (TWR)

Level	T_{on} (μs)	T_{off} (μs)	I (A)
1	0.000753	0.00079	0.00078
2	0.000787	0.00079	0.000787
3	0.000833	0.000793	0.000807
Delta	0.00008	0.000003	0.000027
Rank	1	3	2

Table 6 Response table for S/N ratio for TWR

Level	T_{on} (μs)	T_{off} (μs)	I (A)
1	62.46	62.06	62.17
2	62.08	62.05	62.09
3	61.58	62.02	61.87
Delta	0.88	0.05	0.29
Rank	1	3	2

balance, achieving high Material Removal Rate (MRR) while keeping Tool Wear Rate (TWR) to a minimum.

7 Conclusions

This research successfully demonstrated the application of a circular copper electrode in the EDM of Nimonic C-263. The experimental work involved nine trials designed using the Taguchi method, with Material Removal Rate (MRR) and Tool Wear Rate (TWR) serving as the primary performance indicators. An increase in peak current led to improved MRR, followed by the effects of pulse on-time and off-time. Conversely, TWR was minimized under shorter pulse on-times, with peak current and off-time having secondary influence. Microscopic examination of the machined surface showed the presence of craters and voids, typical of EDM-induced thermal effects. Peak current was the most critical factor for enhancing MRR, while pulse on-time played a major role in determining TWR. The optimal machining configuration consisted of a 10 A current, 7 μs pulse on-time, and 5 μs pulse off-time. These findings highlight the need for careful parameter selection to optimize EDM performance. Future investigations may consider varying electrode materials and geometries, along with exploring advanced optimization strategies, to further improve the machining of high-performance superalloys.

References

1. Manda A, Lakshmanan V, Kuppan P (2016) Wire-electrical discharge machining (WEDM) of Nimonic C-263: optimization of process parameters. J Alloy Compd 662:1–8
2. Amitava M, Pramanik A (2017) Surface integrity of Nimonic C-263 machined by wire electrical discharge machining (WEDM). Mater Manuf Process 32(5):545–552
3. Mouralova K, Benes L, Kolar P (2018) Optimization of powder-mixed EDM parameters using grey relational analysis (GRA) for improved machining performance. J Manuf Process 35:91–99
4. Chekuri R, Desai P, Patel S (2018) Die-sinking EDM optimization using hybrid optimization methods. Procedia CIRP 68:749–754
5. Bhadri R, Rao GV, Sundararajan V (2019) Analysis of machining characteristics using Taguchi method and ANN. J Mech Eng 45(9):1234–1242
6. Chekuri R, Kumar S, Rao PS (2019) Prediction of EDM process parameters for Nimonic C-263 using ANN. Int J Adv Manuf Technol 101(3–4):1023–1032
7. Shastri RK, Mohanty CP (2020) Advanced techniques for cost estimation, micro-hardness, recast layer thickness, and surface crack density in EDM. Mater Today: Proc 27:856–860
8. Pant P, Bharati P (2020) Optimization of EDM process parameters for Nimonic alloys. Eng Sci Technol Int J 23(2):435–442
9. Dewan P, Kundu P (2022) EDM of Nimonic C-263 with a copper tool and deionized water for ecological sustainability. J Clean Prod 264:121393
10. Chekuri R, Desai P, Patel S (2022) Machinability of Nimonic C-263 using die-sinking EDM: Parameter analysis and optimization. Int J Precision Eng Manuf-Green Technol 9(3):401–412

11. Kumar D, Joshi N, Gupta AK (2022) Servo-stabilized electromagnetic levitation system for micro-EDM: enhancing stability and precision. Micromachines 13(1):123
12. Polzer A, Wegener K, Bleicher F (2022) Wire electrical discharge machining (WEDM) of nickel alloys: impact of alloy composition on machinability. J Manuf Process 69:530–540
13. Sahu D, Pramanik A (2022) Surface integrity in powder-mixed EDM of Nimonic C-263. Procedia CIRP 94:167–172
14. Satish K, Rao K, Reddy NS (2022) Optimization of WEDM parameters for Nimonic C-263 using Taguchi's design and MINITAB 17 software. Mater Today: Proc 46:1359–1365
15. Devan P, Kundu P (2023) Optimization of PMEDM parameters for Nimonic C-263 using grey relational analysis. J Mater Process Technol 300:117376
16. Sahu D, Amitava M (2023) Comparison of PMEDM performance under normal and reverse polarity for Nimonic C-263. J Manuf Process 73:123–131
17. Dereje A, Tsegay S, Haftu M (2023) Enhancement of WEDM for Nimonic C-263 using a Taguchi-Grey approach: optimization of MRR, surface roughness, and kerf width. J Clean Prod 312:127803

Applying Machine Learning for Defect Detection in WAAM: A Literature Review

Pawan Kumar, Jitendra Kumar, and Suraj Kumar

Abstract Additive manufacturing (AM) has rapidly advanced, offering unique advantages over traditional manufacturing, and is popular in research and industry. Due to its high deposition rates and cost-effectiveness, Wire Arc Additive Manufacturing is one of the AM processes that works best for making moderately to large-scale metal components. However, WAAM presents key challenges, such as inconsistent surface quality, process parameters, and thermal dynamics that affect dimensional accuracy and overall quality of the fabrication part. These issues complicate the control of geometric parameters like bead width, height, and penetration depth, often leading to deviations from required specifications. This review explores Machine learning (ML) applications in WAAM, which promise to address these production challenges. ML techniques support real-time process optimization, quality control, and predictive modeling in WAAM's complex manufacturing engineering and computational intelligence. ML models provide innovative, data-driven solutions to mitigate variability in WAAM processes, enhancing control and predictability. Machine learning is used for defect and result prediction in additive manufacturing. This study systematically identifies and evaluates key research findings, emphasizing the impact of ML on improving WAAM process outcomes. In conclusion, this review highlights the current challenges and emerging opportunities for advanced ML integration in WAAM.

Keywords Additive manufacturing · WAAM · Machine learning · Artificial intelligence · Defects

P. Kumar (✉) · J. Kumar
Department of Mechanical Engineering, IIT Ropar, Ropar, India
e-mail: Pawan.23mez0006@iitrpr.ac.in

S. Kumar
Department of Chemical Engineering, IIT Madras, Chennai, India

A. Biswas et al. (eds.), *Recent Advancements in Artificial Intelligence and Optimization*, Lecture Notes in Mechanical Engineering,
https://doi.org/10.1007/978-981-95-9268-5_9

1 Introduction

The potential of additive manufacturing (AM) technology, also referred to as its rapid prototyping, 3D printing, or freeform fabrication, to efficiently and precisely build complex components has attracted a lot of attention in recent years. Using CAD software, AM technology slices 3D products into several two-dimensional layers; subsequently, AM, in contrast to conventional manufacturing techniques, drastically streamlines the production process for complex geometries, cuts down on production time, and offers flexible solutions for direct forming and component repair [1]. Although there are several processes under additive manufacturing technology, the categorization of some important AM processes as laser additive manufacturing process [2], powder-based deposition process [3], laminated manufacturing process [4], and wire arc additive manufacturing process [5], as shown in Fig. 1.

Using ballistic impact, metallic and non-metallic particles are mixed and consolidated into a coating or freestanding structure using the cold spray material deposition technique on a suitable substrate [3, 7, 8]. One of the first additive manufacturing processes to be made commercially available was laminated object manufacture (LOM) [4, 9]. It entails cutting and laminating material sheets or rolls layer by layer. A mechanical cutter or laser is used to precisely shape each layer, which is then bonded together in a "form-then-bond" or "bond-then-form" pattern [9]. Material is added to work pieces in additive manufacturing techniques, sometimes in large quantities (as in laser cladding or buildup welding) and other times for the whole production of solid objects. By enabling the conversion of a non-solid source material, such as liquid or powder, into a solid with a predetermined shape, a laser beam frequently plays a critical part in these procedures. The capacity of laser light to be precisely and controllably directed as a focused beam is its main benefit [10]. Wire arc

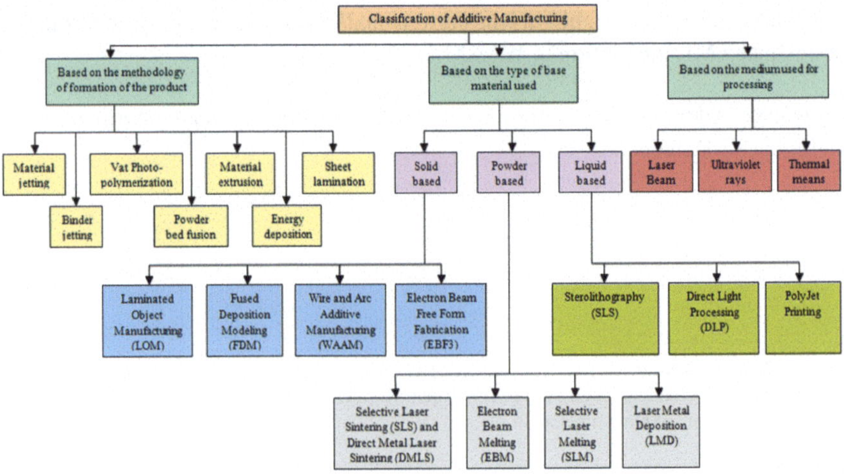

Fig. 1 Types of additive manufacturing process [6]

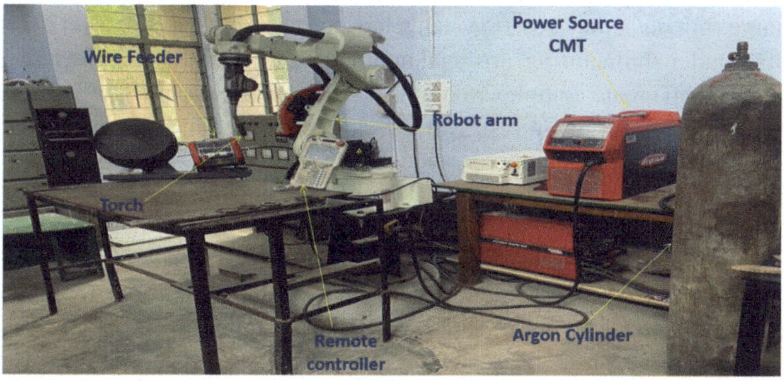

Fig. 2 WAAM setup

additive manufacturing (WAAM) is unique among metal AM methods because of its rapid rate of deposition, high material utilization efficiency, and inexpensive equipment. Because of this, WAAM is especially well-suited for medium to large-scale fabrication of customized metal components. Different types of WAAM techniques include gas metal arc welding (GMAW) [11], and gas tungsten arc welding (GTAW) [12], using the WAAM technique based on plasma arc welding (PAW-WAAM) [13]. A common WAAM technique deposits metal material layer by layer using an electric arc as a heating source [13] to melt the electrode, as shown in Fig. 2, which illustrates the equipment and the WAAM-CMT process with an argon gas cylinder.

These days, the motion mechanism that powers the welding torch is either a high-precision machine tool or an articulated industrial robot. The actuator, welding machine, and other sensors are coordinated by a computer acting as a master controller to carry out the manufacturing operations [14]. The graphs in Fig. 3 illustrate the growing trend in academic research focusing on WAAM and its integration with Machine Learning (ML) over the years and country-wise. The analysis reveals a significant disparity in publication volume between standalone WAAM research (2,795 documents) and its integration with ML (93 documents since 2019). This trend underscores a nascent yet accelerating focus on ML-driven advancements in WAAM [15, 16]. Despite its potential, several challenges prevent WAAM technology from being widely used in manufacturing sectors. The difficulties in simulating the mathematical framework of the WAAM process result from the interconnected elements from different perspectives [15]. previous studies have highlighted how crucial closed-loop control and process monitoring are to improving the reliability, robustness, and repeatability of goods produced using WAAM technologies that are free of defects [17]. Another Challenge of combining WAAM digital models, which deal with different phenomena at different scales, into a single, integrated framework. In this regard, several artificial intelligence (AI) approaches have been recognized as crucial facilitators for tackling these issues in WAAM technology, particularly since they began in 2019 [18]. "In the manufacturing sector, artificial intelligence (AI) methods like ML [19], automation and robotics, machine vision, data mining, big data, and expert systems have proven to be successful [20]. In the world of digital manufacturing, artificial intelligence (AI) approaches are being used more and more to create complex relationships, improve system controllability, and improve product quality as a result of developments in data collecting technologies, robotic systems, and computer science [21]. Traditional WAAM technology, a crucial production method in Industry 4.0, needs to be included in an intelligent fabrication system that integrates modern AI algorithms with sensor networks [22, 23]. Consequently, numerous researchers are employing various AI techniques to enhance WAAM performance through product manufacturing optimization of the design [24, 25], process modeling [26], during operation control, and the examination of defects [27]. Despite significant interest from both academia and industry, recent advancements in ML techniques hold significant potential for enhancing WAAM, particularly in areas such as process monitoring and defect detection. However, there remains a notable research gap in applying these methodologies comprehensively to address critical challenges in WAAM. The integration of advanced ML approaches could improve real-time monitoring, optimize process parameters, and enable early detection of defects, thereby enhancing the reliability and quality of WAAM-fabricated components. Bridging this gap is essential for the wider adoption of WAAM in high-precision industries. Accordingly, this paper aims to provide a systematic review of this area by offering an overview of WAAMs, their integration with Machine Learning (WAAM Plus ML), their scope, and potential research opportunities through an examination of the existing literature. This effort enhances the current understanding of WAAM technologies while guiding future studies toward addressing critical challenges and unlocking new opportunities for innovation. Figure 3 shows the number of publications over the years, as well as the countries involved in this area.

2 Challenges in Implementing Artificial Intelligence and Machine Learning in Manufacturing Technology for WAAM

Wire Arc Additive Manufacturing represents a paradigm shift in additive manufacturing, offering cost-effective solutions for fabricating large-scale metal components with complex geometries [29]. However, the adoption and advancement of WAAM are hampered by several challenges [16, 30]. These obstacles including the productivity of this process, product fabrication geometric accuracy [31], and quality assurance, impede its full-scale implementation in industrial applications [32]. Nowadays, various full products are manufactured by WAAM. Figure 4a and b show that the propeller and water turbine blades were manufactured by wire arc additive manufacturing, respectively [33].

Wire Arc Additive Manufacturing (WAAM) offers significant productivity advantages, with high material deposition rates of 3–5 kg/hour for ferrite-based materials and 1–2 kg/hour for aluminum alloys. But, achieving these rates is challenging

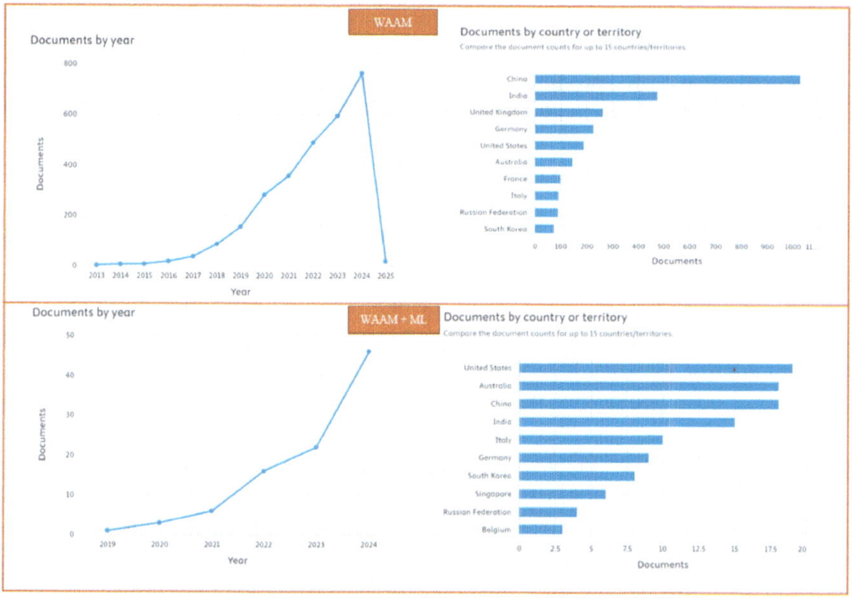

Fig. 3 Year vs. Publication Trend & Right-Side Country Comparison [28]

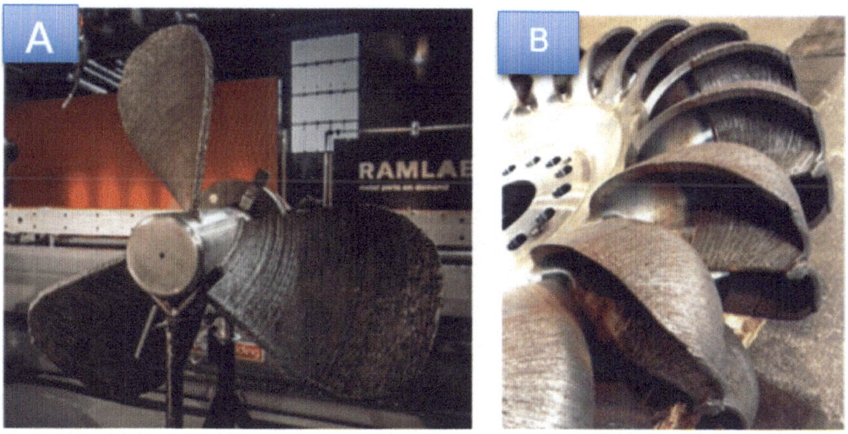

Fig. 4 WAAM part **(a)** propeller and **(b)** water turbine blades [32]

for workpieces with complex geometries or unique features [5]. A key issue is the absence of a universal programming strategy for diverse geometries and materials, leading to time-intensive custom programming and time-consuming testing. Also, prolonged defect-free operation is difficult without constant operator intervention, reducing productivity and increasing costs. Addressing these challenges requires a multi-sensor system capable of real-time monitoring and autonomous defect correction to improve efficiency and reliability. When talk about structural accuracy or complex geometry, the product's fabricated precision of the WAAM process is primarily influenced by individual weld beads of layers [34]. For single-layer or bead deposition, the inherent limitations of arc welding restrict precision to the millimeter scale, as feeding wire sizes vary from 0.9 to 1.8 mm. Errors in bead placement or overlap, along with arc ignition and extinction, introduce defects in the fabricated parts. For overall displacement, negligible per-layer errors can accumulate across layers, causing significant deviations in the final part. Additionally, heat accumulation distorts substrates and workpieces, further reducing deposition accuracy [35]. To mitigate these challenges, researchers have explored predictive modeling and real-time control systems. Neural network and support vector regression methods predict weld bead geometries but often rely on static data, limiting adaptability to dynamic WAAM processes. Advancements like model-based reinforcement learning frameworks and layer-by-layer controllers show promise but require further development to enhance automation, robustness, and real-time error correction capabilities [35, 36]. The adoption of WAAM in aerospace and maritime industries demands the production of metal components that adhere to rigorous quality standards, as defects can result in catastrophic failures [37]. The prolonged and complex thermal and material transfer processes in WAAM often lead to defects such as porosity, cracks, and voids, which are closely linked to process signals like electrical, optical, and acoustic signals. Real-time monitoring and defect control are, therefore, essential. Innovations like the method for detecting irregularities in real time by Cho et al. [38], using convolutional neural networks. These systems detect defects such as balling and bead-cut issues during deposition. However, challenges remain, including slow response times, limited defect detection capabilities, and inefficiencies in data collection [39]. Developing a robust system capable of autonomously identifying and correcting defects in real time is critical for ensuring WAAM quality. Virtualization in manufacturing involves creating virtual environments that simulate real-world conditions, hardware functions, and signal feedback. In WAAM, virtualization facilitates real-time interactions and updates via cyber-physical systems and sensors, enhancing operator safety, improving knowledge transfer, and enabling efficient production data collection [40]. Virtualized environments, such as Wang et al.'s virtual welding system, allow operators to remotely observe and control processes using technologies like virtual reality (VR). Despite its benefits, WAAM virtualization is hindered by challenges in integrating diverse platform technologies, sensors, and systems, as well as gaps in understanding virtualization frameworks and methodologies [41]. Similarly, Tang et al. applied deep learning algorithms to classify WAAM surface images and identify defects. While enabling technologies such as advanced sensing, digital signal processing, simulation, and AI provide promising solutions, significant advancements in integration and system intelligence are necessary to overcome productivity, accuracy, quality assurance, and virtualization challenges, Integrating AI with Wire Arc Additive Manufacturing requires substantial computational resources due to the complexity of AI algorithms. The need to process large datasets and enable real-time analysis demands high-performance computing. Additionally, algorithm validation is critical, ensuring that AI-driven solutions operate reliably and consistently in industrial-scale environments [42]. Without thorough validation, the risk of AI solutions failing to meet quality standards or adapt to variable manufacturing conditions could hinder their adoption in practical applications. Both computational power and robust validation are essential for the successful integration and scalability of AI in WAAM processes.

3 Machine Learning Applications in WAAM Defect Detection

The key objectives of employing AI in real-time monitoring and defect detection are defect classification, defect recognition, process parameter optimization, and predictive maintenance. Defect classification involves identifying specific defect types, such as cracks, voids, or porosity, while defect recognition focuses on distinguishing normal process variations from actual defects. Parameter optimization ensures that process variables, such as temperature, speed, and feed rate, are automatically adjusted to minimize defects, thereby enhancing production efficiency. Predictive maintenance, another critical objective, involves using AI to predict the likelihood of defects before they occur, allowing operators to take preventive actions and avoid costly downtime. Table 1 presents a literature review on AI-integrated WAAM systems, emphasizing the input, output, and data-driven models used.

4 The Importance of Intelligent WAAM Systems

4.1 Real-Time Monitoring

This apprpach uses sensors and data acquisition systems to monitor key parameters such as temperature, voltage, current [67], and deposition rates [68]. During the WAAM process, the Feedback Mechanism adjusts process parameters dynamically based on real-time data to maintain optimal performance and minimize defects.

4.2 Defect Prediction

Employs ML models to predict defects like porosity, cracks, or uneven deposition, enabling early action. Performance Forecasting involving predicting material performance and mechanical properties of the final part based on processing conditions. Utilized ML models, including k-NN, random forest, and SVR, to predict elongation

Table 1 Literature review of data-driven models in WAAM: methods, inputs, and outputs

Type of information provided	Data-driven models	Research output	Refs.
Time–frequency domain (Fourier and wavelet transforms)	Computer vision-enhanced feature extraction	Feature extraction for WAAM defect monitoring	[43]
Process monitoring signals	Particle-swarm-optimization-assisted deep transfer learning	Detecting outliers and anomalies in additive manufacturing	[44]
Integrated visual and time–frequency data	Deep learning with computer vision	Advanced defect monitoring in WAAM processes	[45]
Simulation data of the 3D weld pool surface	LSTM(Long Short-Term Memory)	Real-time 3D weld pool prediction	[46]
Welding parameters, Electrical signals, sound signals (MFCC)	BiLSTM-CTC	Errors detected, 0.295 classification error rate	[47]
Magnetic flux leakage (MFL) signals (time–frequency graph)	Generative Adversarial Network (GAN) with Auxiliary Classification	99.5% defect recognition accuracy, faster training	[48]
Eddy current signals (time–frequency diagram), 3D laser scan images	CNN, VGG16	Enhanced defect identification accuracy and real-time detection efficiency	[49]
Visual and acoustic signals from a dual-sensor system	Hybrid CNN and ELM for feature extraction and classification	Penetration classification with superior accuracy for the VPPAW process	[50]
Magneto-optical images of low-carbon steel WAAM products	Cascaded Convolutional Neural Networks (CSCNN)	Automatic detection of surface defects in WAAM products	[51]
Arc spectrum, sound, and voltage signals from the GTAW process	A hybrid Fisher-based filter and wrapper for feature selection, followed by a classification model	94.72% accuracy in detecting welding defects for Al alloy in GTAW	[52]
WAAM defect dataset, images from the manufacturing process	YOLO-attention based on YOLOv4	Automatic defect detection	[53]
Ultrasonic signals are converted into 2D spectrograms	CNN models—ResNet18, esSNet50, ResNet101, Darknet19, modified Darknet19	Modified Darknet19 achieves 66.69% accuracy, with sensitivity of 57.14%, 60%, and 100% for three defects	[20]

(continued)

Table 1 (continued)

Type of information provided	Data-driven models	Research output	Refs.
Ultrasonic defect echo signals, wavelet coefficients	SVM,1D-LBP	Multi-class defect classification,98.3% defect classification accuracy	[54]
Welding arc data (voltage, arc length, bead geometry, etc.)	Random forest	Predictive algorithms for defect detection in WAAM layers	[55]
Reflected laser and visible light radiation signals	Convolutional Neural Networks	In situ judgment of laser welding quality	[56]
Welding arc current and voltage signals	Support Vector Machine (SVM) for defect detection	Detection of WAAM process defects with over 90% F1-score	[57]
Welding parameters (e.g., thickness, width)	Conditional Generative Adversarial Network (cGAN)	Prediction of deformation field with high accuracy, validated by comparison with the numerical method	[58]
Coherent optical diagnosis signals during laser welding	Deep Belief Network (DBN) for online identification	In situ Identification of porosity regions during laser welding	[59]
Welding signals (keyhole, plume, spatter, etc.)	LDA, Stacked Auto-Encoder (SAE)	Online welding status monitoring and defect classification	[60]
UVV visual sensor, spectrometer, photodiode, auxiliary illumination signals	Deep Belief Network (DBN), Genetic Algorithm	Accurate, robust monitoring of welding status	[61]
Multiple-sensor signals (keyhole, plume, spatter, optical, and spectrum information)	Correlation analysis, Linear Discriminant Analysis (LDA), Stacked Auto-Encoder (SAE)	Online monitoring of welding statuses, detection of blowout, humping, and undercut defects	[62]
High-dimensional spectrometer signals, welding parameters, and real-time welding data	Stacked Auto-Encoder (SAE), Softmax Classifier, Particle Swarm Optimization (PSO)	Improved welding defect detection accuracy, real-time status monitoring	[63]
Acoustic signals from the weld site	SeCNN-LSTM model, compared with CNN-SVM,	Weld flaw detection accuracy of 91%	[64]

(continued)

Table 1 (continued)

Type of information provided	Data-driven models	Research output	Refs.
Digital X-ray images of welding defects	Improved U-Net network, data augmentation	Welding defect detection precision of 88.4%	[65]
Calibration data, reference point positioning data, and tracking trajectory data	Generative Adversarial Network (GAN)	Improved calibration efficiency, low uncertainty in point positioning, robust tracking accuracy	[66]

and strain in WAAM specimens with high accuracy, achieving an R^2 of 99.97%.ML in reducing experimental costs and optimizing mechanical property predictions [69].

4.3 Path Planning

Optimizes tool paths to reduce waste, improve deposition accuracy, and enhance surface finish. The authors proposed planner, a reinforcement learning-based framework for automatic deposition path planning in WAAM. By incorporating sequential least squares programming and automated preprocessing, it optimizes path adaptability to thin-walled structures while enabling real-time adjustments to welding speed wire and feed rate for enhanced geometry conformity [70].

4.4 Material Intelligence

This approach leverages data-driven insights to recommend suitable alloys or material blends based on desired properties while ensuring uniform material distribution and high deposition quality through real-time monitoring of melt pool dynamics and bead geometry. The study addresses irregularities in deposition bead geometry, particularly in the arc strike zone, without altering the deposition path or requiring external devices. Using a support vector machine (SVM) classifier, deposition condition ranges were determined for achieving uniform bead shapes. SVM regression further optimized conditions to minimize deviations between arc strike and deposition zones. Validation experiments involving five-layer deposition demonstrated reduced average height and width errors in the arc strike zone (0.36 and 1.28%) and middle deposition zone (0.64% and 1.26%). The results confirm the regression model's effectiveness in enhancing deposition accuracy and reducing post-processing costs [71]. This study developed a SVM classifier to optimize bead geometry in WAAM, effectively predicting optimal deposition conditions with minimal input data. Verification experiments confirmed its efficiency, demonstrating potential for real-world applications in remanufacturing and multilayer deposition process [72].

4.5 Fault Detection and Diagnosis

Anomaly detection in manufacturing involves identifying and resolving irregularities within the production process using pattern recognition and ML models. The author presents a novel ML framework that quantitatively analyzes the relationship between WAAM process parameters and deposition shape, offering an optimal process parameter selection to control final geometry. It aims to improve prediction accuracy and process reliability, with experimental validation for the proposed method [67].

5 Challenges in AI-Powered Real-Time Defect Detection for WAAM

5.1 Limitations of Sensing Devices

The sensors used in WAAM, such as microphones, ultrasonic probes, X-ray detectors, and optical cameras, face multiple limitations during real-time monitoring. The intense wide-band arc light generated during welding interferes with the performance of optical devices, while ultrasonic and acoustic sensors experience signal distortion due to noise.

5.2 Data Overload and Computational Burden

The high-volume sensor data from WAAM strains computational resources. Techniques such as dimensionality reduction (PCA, autoencoders), edge/cloud computing, and digital twins optimize processing by compressing data, offloading tasks, and virtualizing real-time monitoring.

5.3 Defect Classification and Recognition

Classifying and recognizing specific types of defects in WAAM can be challenging due to the wide variety of potential defects, such as cracks, pores, surface roughness, and voids. Compounding this issue is the lack of clearly defined thresholds for classifying these defects, making it difficult to set uniform quality standards. Traditional data-driven models often struggle to identify subtle or uncommon defects, leading to misclassification. Solutions to these challenges include the use of physics-driven deep learning models that combine physical simulations with ML.

6 Future Research Opportunities

6.1 *Digital Twin*

A digital twin (DT) is characterized as a dynamic, multi-physics, and multi-scale virtual representation of a physical system or component, built using high-fidelity physical models to mirror real-world behavior [139]. In the context of WAAM, the primary objectives of DTs are to facilitate process simulation, real-time monitoring, and autonomous control of manufacturing systems and products [73]. The development of DTs for WAAM requires the integration of three fundamental elements: process modeling, sensor-driven data acquisition, and adaptive control mechanisms. For instance, AI algorithms can simulate process behavior to determine optimal parameter settings that ensure superior product quality. Moreover, AI-driven process tracking systems provide continuous feedback, enabling autonomous corrective actions when deviations from desired performance are detected [41].

6.2 *Human–robot Collaboration*

To combine the high movement accuracy of industrial robots with the intelligent adaptations of human welders, researchers have recently suggested and implemented human–robot collaboration manufacturing systems in welding domains. For the system to accomplish common objectives in the welding process, such as increasing fabrication efficiency and accuracy, human and robot agents need to work together. The WAAM process may face more challenging scenarios in strategy creation, parameter selection, and deposition control than the welding process, where innovative human–robot collaborative techniques are required to satisfy the objectives of accuracy, efficiency, and cost. Professionally trained human operators could contribute in certain important ways to improving the WAAM process's performance [74].

6.3 Imitation Learning

One goal of using AI approaches in WAAM systems is to fully automate the process and reduce the human need. To do this, the system must possess specific decision-making and self-learning capabilities. WAAM systems may be able to acquire the required capabilities through imitation learning, which has been used in industries to assist robots in doing novel jobs. Imitation learning, as opposed to human–robot collaboration, seeks to learn from patterns given by human teachers and use the acquired information to do practical tasks with judgmental support. Real production can confirm the accuracy of the knowledge and experience that have been learnt or extracted, and the manufacturing system may be continuously adjusted to ensure performance. To help WAAM systems understand how a professional human makes judgements, imitation learning could be used for several WAAM process steps, including modeling, path planning, and material deposition [17].

7 Conclusions

Wire Arc Additive Manufacturing (WAAM) has a lot of potential for building large metal parts quickly and affordably. However, it still faces major challenges like defects, poor accuracy, and difficulty in controlling the process. Recent research shows that applying Machine learning (ML) techniques can help solve many of these problems. Machine learning helps in monitoring the process in real-time, predicting defects early, optimizing tool paths, and improving the quality of the final parts. Especially models like Deep Belief Networks (DBN) and Convolutional Neural Networks (CNN) have shown good results in detecting defects and enhancing WAAM reliability. Technologies such as digital twins, human–robot collaboration, and imitation learning have the potential to make WAAM even smarter and more efficient in the future. In summary, integrating WAAM with advanced machine learning tools presents a promising approach to achieving smarter, faster, and more reliable metal manufacturing system.

References

1. Ngo TD, Kashani A, Imbalzano G, Nguyen KT, Hui D (2018) Additive manufacturing (3D printing): A review of materials, methods, applications, and challenges. Compos B Eng 143:172–196
2. Lee H, Lim CHJ, Low MJ, Tham N, Murukeshan VM, Kim Y-J (2017) Lasers in additive manufacturing: A review. Int J Precis Eng Manuf-Green Technol 2017, 4:307 22
3. Ladani L, Sadeghilaridjani M (2021) Review of powder bed fusion additive manufacturing for metals. Metals 11(9):1391
4. Feygin M, Hsieh B (1991) Laminated object manufacturing (LOM): a simpler process

5. Li Y, Su C, Zhu J (2022) Comprehensive review of wire arc additive manufacturing: Hardware system, physical process, monitoring, property characterization, application and future prospects. Results Eng 13:100330
6. Saleh Alghamdi S, John S, Roy Choudhury N, Dutta NK (2021) Additive manufacturing of polymer materials: Progress, promise and challenges. Polymers 13(5):753
7. Assadi H, Gärtner F, Stoltenhoff T, Kreye H (2003) Bonding mechanism in cold gas spraying. Acta Mater 51(15):4379–4394
8. Cappelli G, Sasnauskas A, Yin S, Lupoi R (2024) Erosion behavior of cold-sprayed coatings made of CoCrFeMnNi HEA and tungsten carbide nanoparticles in a Nickel Matrix. J Therm Spray Technol 33(2):471–488
9. Gu DD, Meiners W, Wissenbach K, Poprawe R (2012) Laser additive manufacturing of metallic components: materials, processes and mechanisms. Int Mater Rev 57(3):133–164
10. Pattanayak S, Sahoo SK (2021) Gas metal arc welding based additive manufacturing—a review. CIRP J Manuf Sci Technol 33:398–442
11. Geng H, Li J, Xiong J, Lin X, Zhang F (2017) Optimization of wire feed for GTAW based additive manufacturing. J Mater Process Technol 243:40–47
12. Artaza T, Suárez A, Veiga F, Braceras I, Tabernero I, Larrañaga O, Lamikiz A (2020) Wire arc additive manufacturing Ti6Al4V aeronautical parts using plasma arc welding: Analysis of heat-treatment processes in different atmospheres. J Mater Res Technol 9(6):15454–15466
13. Pires JN, Loureiro A, Bölmsjo G (2006) Welding robots: technology, system issues and application. Springer Science & Business Media
14. Scopus—Document search n.d. https://www.scopus.com/search/form.uri?display=basic#basic (Accessed Dec, 2024)
15. Thomas-Seale LE, Kirkman-Brown JC, Attallah MM, Espino DM, Shepherd DE (2018) The barriers to the progression of additive manufacture: Perspectives from UK industry. Int J Prod Econ 198:104–118
16. He F, Yuan L, Mu H, Ros M, Ding D, Pan Z, Li H (2023) Research and application of artificial intelligence techniques for wire arc additive manufacturing: a state-of-the-art review. Robot Comput-Integr Manuf, 82, 102525)
17. He F, Yuan L, Mu H, Ros M, Ding D, Pan Z, Li H (2023) Research and application of artificial intelligence techniques for wire arc additive manufacturing: a state-of-the-art review. Robot Comput-Integr Manuf, 82, 102525
18. Sarker IH (2021) Machine learning: Algorithms, real-world applications and research directions. SN Comput Sci 2(3):160
19. Knaak C, Kolter G, Schulze F, Kröger M, Abels P (2019) Deep learning-based semantic segmentation for in-process monitoring in laser welding applications. In Appl Mach Learn. 11139, pp 10–22. SPIE
20. Li W, Zhang H, Wang G, Xiong G, Zhao M, Li G, Li R (2023) Deep learning based online metallic surface defect detection method for wire and arc additive manufacturing. Robot Comput-Integr Manuf 80:102470
21. Kumar S, Gopi T, Harikeerthana N, Gupta MK, Gaur V, Krolczyk GM, Wu C (2023) Machine learning techniques in additive manufacturing: a state of the art review on design, processes and production control. J Intell Manuf 34(1):21–55
22. Ahmmed MS, Isanaka SP, Liou F (2024) Promoting synergies to improve manufacturing efficiency in industrial material processing: A systematic review of industry 4.0 and AI. Machines, 12(10), 681
23. Alfaro SC, Franco FD (2010) Exploring infrared sensoring for real time welding defects monitoring in GTAW. Sensors 10(6):5962–5974
24. Morris PT, Cavender DP (2018) Additively manufactured thermal standoff and injector for a 440N low-toxicity monopropellant thruster. In: 2018 joint propulsion conference (p 4895)
25. Liu L, Chen H, Chen S (2022) Quality analysis of CMT lap welding based on welding electronic parameters and welding sound. J Manuf Process 74:1–13
26. Rohe M, Stoll BN, Hildebrand J, Reimann J, Bergmann JP (2021) Detecting process anomalies in the GMAW process by acoustic sensing with a convolutional neural network (CNN) for classification. J Manuf Mater Process 5(4):135

27. Brennan MC, Keist JS, Palmer TA (2021) Defects in metal additive manufacturing processes
28. Scopus—Document search results n.d. https://www.scopus.com/results/results.uri?sort=plf-f&src=s&st1=waam&sid=a3fc8a2fdcc228417309a548d6ed4924&sot=b&sdt=b&sl=19&s=TITLE-ABS-KEY%28waam%29&origin=searchbasic&editSaveSearch=&yearFrom=Before+1960&yearTo=Present&sessionSearchId=a3fc8a2fdcc228417309a548d6ed4924&limit=10 (Accessed December 8, 2024)
29. Liu J, Xu Y, Ge Y, Hou Z, Chen S (2020) Wire and arc additive manufacturing of metal components: a review of recent research developments. Int J Adv Manuf Technol 111(1):149–198
30. Kawalkar R, Dubey HK, Lokhande SP (2022) Wire arc additive manufacturing: A brief review on advancements in addressing industrial challenges incurred with processing metallic alloys. Materials today: proceedings 50:1971–1978
31. Zahidin MR, Yusof F, Rashid SHA, Mansor S, Raja S, Jamaludin MF, Hussein NIS (2023) Research challenges, quality control and monitoring strategy for Wire Arc Additive Manufacturing. J Mater Res Technol 24:2769–2794
32. Pant H, Arora A, Gopakumar GS, Chadha U, Saeidi A, Patterson AE (2023) Applications of wire arc additive manufacturing (WAAM) for aerospace component manufacturing. Int J Adv Manuf Technol 127(11–12):4995–5011
33. Treutler K, Wesling V (2021) The current state of research of wire arc additive manufacturing (WAAM): a review. Appl Sci 11(18):8619
34. Gor M, Soni H, Rajput GS, Shah H, Sahlot P (2021) A review on wire arc additive manufacturing: effect of process parameters on the build material properties. Adv Eng Mater: Sel Proc FLAME 2020:247–255
35. Xie D, Lv F, Liang H, Shen L, Tian Z, Zhao J, Shuai C (2021) Towards a comprehensive understanding of distortion in additive manufacturing based on assumption of constraining force. Virtual Phys Prototyp 16(sup1):S85–S97
36. Tran HS, Aizpuru N, Relich P, Pichay K, Ponniah S (2021) Real-time optical defect detection, identification and correction technology for wire-feed additive manufacturing. In SAMPE neXus 2021. Society for the Advancement of Material and Process Engineering–North America.
37. Zhang Z, Wen G, Chen S (2017) Audible sound-based intelligent evaluation for aluminum alloy in robotic pulsed GTAW: mechanism, feature selection, and defect detection. IEEE Trans Industr Inf 14(7):2973–2983
38. Huang C, Wang G, Song H, Li R, Zhang H (2022) Rapid surface defects detection in wire and arc additive manufacturing based on laser profilometer. Measurement 189:110503
39. Wang J, Mu C, Mu S, Zhu R, Yu H (2023) Welding seam detection and location: Deep learning network-based approach. Int J Press Vessels Pip 202:104893
40. Zimermann R, Mohseni E, Vasilev M, Loukas C, Vithanage RK, Macleod CN, Ding J (2022) Collaborative robotic wire+ arc additive manufacture and sensor-enabled in-process ultrasonic non-destructive evaluation. Sensors 22(11):4203
41. Wang Y, Xu X, Zhao Z, Deng W, Han J, Bai L, Yao J (2021) Coordinated monitoring and control method of deposited layer width and reinforcement in WAAM process. J Manuf Process 71:306–316
42. Soori M, Jough FKG, Dastres R, Arezoo B (2024) AI-based decision support systems in Industry 4.0, A review. J Econ Technol
43. Mattera G, Polden J, Nele L (2024) A time-frequency domain feature extraction approach enhanced by computer vision for wire arc additive manufacturing monitoring using fourier and wavelet transform. J Adv Manuf Syst
44. Fang J, Wang Z, Liu W, Lauria S, Zeng N, Prieto C, Liu X (2022) A new particle swarm optimization algorithm for outlier detection: Industrial data clustering in wire arc additive manufacturing. IEEE Trans Autom Sci Eng 21(2):1244–1257
45. Mu H, Chen Z, He F, Li Y, Xia C, Commins P, Pan Z (2020) Defect detection and process monitoring for wire arc additive manufacturing using machine learning. In international conference on robotic welding, intelligence and automation (pp. 3–22). Singapore: Springer Nature Singapore

46. Li L, Cheng F, Wu S (2021) An LSTM-based measurement method of 3D weld pool surface in GTAW. Measurement 171:108809
47. Zeng Y, Wang X, Qin X, Hua L, Xu M (2021) Laser Ultrasonic inspection of a Wire+ Arc Additive Manufactured (WAAM) sample with artificial defects. Ultrasonics 110:106273
48. Lang X (2022) Recognition method of pipeline weld defects based on auxiliary classifier generative adversarial networks. IEEE Instrum Meas Mag 25(2):69–77
49. Miao R, Shan Z, Zhou Q, Wu Y, Ge L, Zhang J, Hu H (2022) Real-time defect identification of narrow overlap welds and application based on convolutional neural networks. J Manuf Syst 62:800–810
50. Droubi MG, Faisal NH, Orr F, Steel JA, El-Shaib M (2017) Acoustic emission method for defect detection and identification in carbon steel welded joints. J Constr Steel Res 134:28–37
51. Zhu L, Luo Y, Han J, Zhang C, Xu J, Chen D (2019) Energy characteristics of droplet transfer in wire-arc additive manufacturing based on the analysis of arc signals. Measurement 134:804–813
52. Zhang Z, Chen H, Xu Y, Zhong J, Lv N, Chen S (2015) Multisensor-based real-time quality monitoring by means of feature extraction, selection and modeling for Al alloy in arc welding. Mech Syst Signal Process 60:151–165
53. Zhang Z, Wen G, Chen S (2019) Weld image deep learning-based on-line defects detection using convolutional neural networks for Al alloy in robotic arc welding. J Manuf Process 45:208–216
54. Hu H, Peng G, Wang X, Zhou Z (2018) Weld defect classification using 1-D LBP feature extraction of ultrasonic signals. Nondestruct Test Eval 33(1):92–108
55. Cheepu M (2023) Machine learning approach for the prediction of defect characteristics in wire arc additive manufacturing. Trans Indian Inst Met 76(2):447–455
56. Zhang Y, Gao X, You D, Jiang X (2022) Online judgment of laser welding process based on time-frequency analyses of in-situ high-speed optical signals. Optik 265:169552
57. Li Y, Polden J, Pan Z, Cui J, Xia C, He F, Wang L (2022) A defect detection system for wire arc additive manufacturing using incremental learning. J Ind Inf Integr 27:100291
58. Guo R, Liu H, Xie G, Zhang Y (2021) Weld defect detection from imbalanced radiographic images based on contrast enhancement conditional generative adversarial network and transfer learning. IEEE Sens J 21(9):10844–10853
59. Ma D, Jiang P, Shu L, Qiu Y, Zhang Y, Geng S (2023) DBN-based online identification of porosity regions during laser welding of aluminum alloys using coherent optical diagnosis. Opt Laser Technol 165:109597
60. Ribas LC, Casaca W, Fares RT (2025) Conditional generative adversarial networks and deep learning data augmentation: a multi-perspective data-driven survey across multiple application fields and classification architectures. AI, 6(2), 32
61. Liu M, Li Y, Wang Z (2021) A propelled multiple fusion deep belief network for weld defects detection. In: Proceedings of the 2021 2nd international conference on control, robotics and intelligent system (pp. 141–146)
62. Zhang Y, Zhang N, You D, Gao X, Katayama S (2019) High-power disk laser welding statuses monitoring based on analyses of multiple-sensor signals. J Manuf Process 41:221–230
63. Zhang Y, Gao X, You D, Zhang N (2019) Data-driven detection of laser welding defects based on real-time spectrometer signals. IEEE Sens J 19(20):9364–9373
64. Ji T, Mohamad Nor N (2023) Deep learning-empowered digital twin using acoustic signal for welding quality inspection. Sensors 23(5):2643
65. Yang L, Wang H, Huo B, Li F, Liu Y (2021) An automatic welding defect location algorithm based on deep learning. NDT and E Int 120:102435
66. Lee H, Sung K, Park H, Rhee SH (2004) Measurement of weld bead defect for shock absorber using laser vision sensor. Key Eng Mater 270:2332–2337
67. Xiao X, Waddell C, Hamilton C, Xiao H (2022) Quality prediction and control in wire arc additive manufacturing via novel machine learning framework. Micromachines 13(1):137
68. Xia C, Pan Z, Polden J, Li H, Xu Y, Chen S (2022) Modelling and prediction of surface roughness in wire arc additive manufacturing using machine learning. J Intell Manuf 33(5):1467–1482

69. Chigilipalli BK, Veeramani A (2025) A machine learning approach for the prediction of tensile deformation behavior in wire arc additive manufacturing. Int J Interact Des Manuf (IJIDeM) 19(1):185–197
70. Petrik J, Bambach M (2023) Reinforcement learning and optimization-based path planning for thin-walled structures in wire arc additive manufacturing. J Manuf Process 93:75–89
71. Oh WJ, Lee CM, Kim DH (2022) Prediction of deposition bead geometry in wire arc additive manufacturing using machine learning. J Mater Res Technol 20:4283–4296
72. Kim DO, Lee CM, Kim DH (2024) Determining optimal bead central angle by applying machine learning to wire arc additive manufacturing (WAAM). Heliyon, 10(1)
73. Nele L, Mattera G, Yap EW, Vozza M, Vespoli S (2024) Towards the application of machine learning in digital twin technology: a multi-scale review. Discov Appl Sci 6(10):502
74. Li S, Zheng P, Liu S, Wang Z, Wang XV, Zheng L, Wang L (2023) Proactive human–robot collaboration: Mutual-cognitive, predictable, and self-organising perspectives. Robot Comput-Integr Manuf 81:102510

Modeling and Finite Element Analysis of the Combustion Chamber in the Motor Mount of the First Stage of a Missile

Sivam Karmakar, Muskan Borgohain, Velpula Tharun, Sushen Kirtania, Sanjib Banerjee, and Satadru Kashyap

Abstract In the present work the combustion chamber of the first stage of a missile has been modeled, which is cylindrical in shape. The cylindrical combustion chamber has been modeled inside the motor mount composed of graphite fiber in epoxy matrix with multi-directional fiber orientation. In the present analysis three types of materials, namely, titanium alloy, laminated composites with fiber orientations [0°/ ± 45°/90°]s and [0°/ ± 30°/ ± 45°/90°]s have been considered for finite element modeling and analysis of combustion chamber of a missile. A structural analysis of the combustion chamber has been conducted for the three types of materials to evaluate its performance under operating conditions involving total deformation and hoop stress. This generated stresses due to mechanical loads have been analyzed to verify the maximum limit that can be withstand by the specified materials. This step is essential for validating the design of the cylindrical chamber can withstand the combined effects of mechanical loads and operational conditions without failure. It has been found from the finite element analysis that the stress produced due to mechanical load is highest in the titanium alloy and least in the composite with fiber orientation [0°/ ± 30°/ ± 45°/90°]s, showing the least deformation in the composite with fiber orientation [0°/ ± 30°/ ± 45°/90°]s, followed by the composite with fiber orientation [0°/ ± 45°/90°]s and showing the highest deformation in the titanium alloy. Therefore, it could be concluded that the composite with fiber orientation [0°/ ± 30°/ ± 45°/90°]s is suitable material for missile combustion chamber.

Keywords Deformation · Fiber orientation · Finite element analysis · Graphite/ epoxy composite · Hoop stress

S. Karmakar · M. Borgohain (✉) · V. Tharun · S. Kirtania · S. Banerjee · S. Kashyap
Department of Mechanical Engineering, Tezpur University, Tezpur, Assam, India
e-mail: muskanborgohain2102@gmail.com

A. Biswas et al. (eds.), *Recent Advancements in Artificial Intelligence and Optimization*, Lecture Notes in Mechanical Engineering,
https://doi.org/10.1007/978-981-95-9268-5_10

1 Introduction

In recent times, there has been a noticeable increase in the use of composite materials in aerospace vehicles [1]. One of the most used composites is carbon fiber (CF) epoxy composites, primarily CF-based reinforced polymer (CFRP) [2] and graphite fiber (GF) epoxy composite [3]. The advantages of using GF epoxy composites in space applications are significant. These composites provide lightweight solutions with high structural performance, crucial for spacecraft missions that demand lightweight structures with high pointing accuracy and stability in dynamic and thermal environments. Lukez and Thiokol [4] explored the feasibility and benefits of utilizing GF epoxy composites in spacecraft structures. These composites offer high specific stiffness, low coefficient of thermal expansion (CTE), and exceptional dimensional stability, making them ideal for spacecraft structures. Lakshmi and Rao [5] presented a review on how GF epoxy composites offer benefits like high modulus, strength, and thermal stability, making them suitable for withstanding extreme conditions encountered in space, including vacuum, thermal radiation, atomic oxygen, ionizing radiation, and plasma. Nath et al. [6] have modeled an aeroplane wing out of Titanium alloy and CFRP with resin epoxy with varying ply orientation. Then they have analyzed the different weight to strength ratios along with the deformation and stresses experienced to study and compare the strength of the materials. Their analysis shows how the use of composites to replace the traditional materials yields better strength-to-weight ratio and a higher tolerance to stresses and deformation. Jamison et al. [7] discussed how composite hardware can be engineered and manufactured to lower the number of parts and shorten assembly time. They also reported how the ability to mold components close to their final shape minimizes the need for additional machining. This approach helped to streamline the production process and enhances efficiency in the manufacturing of missile components. Yasaka and Onoda [8] have stated how GF epoxy composite materials are increasingly utilized for both the primary and the secondary structures to take advantage of the superior mechanical properties of the material. In GF epoxy, material properties can be tailored, by selecting orientations and contents of GF, for strength, rigidity, and thermal expansion.

From the above discussion, it is evident that many researchers are working to develop new composites to replace conventional materials in aerospace parts. To promote greater strength-to-weight ratio, traditional metals are being replaced by graphite/epoxy composites in areas requiring high structural and thermal resistance. However, there is limited investigation into the impact of analyzing the use of different fiber orientation of composite materials in designing aerospace vehicles. This study aims to provide a comprehensive structural analysis using finite element method (FEM) to evaluate the behavior of combustion chamber in the motor mounts under extreme pressure and temperatures, ensuring safety and compliance with design limits.

2 Material and Methodology

2.1 Material Selection for Combustion Chamber

Based on the advanced mechanical properties [9], a graphite/epoxy composite has been considered for the manufacture of combustion chamber in the first stage of a prototype missile. Graphite/epoxy is a composite material made of GF embedded in an epoxy matrix. The GF provide strength and rigidity, while epoxy resins are used as matrices because they are easy to process, have good thermal characteristics, and have low shrinkage during cure. The properties of graphite/epoxy composite are listed in Table 1.

From the above data, the rule of mixture (ROM) method [10] has been applied to find out the volume percentage of fiber as well as that of matrix as

$$\frac{1}{E_2} = \frac{V_f}{E_f} + \frac{V_m}{E_m} \tag{1}$$

where *E2, Ef,* and *Em* are Young's modulus of the composite in transverse direction, Young's modulus of the fiber, and matrix material, respectively, and *Vf* and *Vm* are the volume fraction of the fiber and matrix materials, respectively.

Therefore, by applying (1) for the given values of Young's modulus of the fiber and matrix and for the given value of transverse modulus of the graphite epoxy composite, we can find the volume percentage of fiber as 64 wt.% and that of matrix as 36 wt.%.

The titanium alloy can arguably be called one of the traditional materials used in the aerospace sector [11]. For analysis purpose, the titanium alloy properties are considered from the ANSYS software with a density of 4,430 kg/m3, Young's modulus of 110 GPa, Poisson's ratio of 0.33, tensile strength of 950 MPa, and yield

Table 1 Properties of graphite epoxy composite [5]

Properties of graphite epoxy	Values
Longitudinal modulus	140.68 GPa
Transverse modulus	9.13 GPa
In plane shear modulus	7.24 GPa
Poisson ratio	0.30
Longitudinal tensile strength	1.447 GPa
Longitudinal compressive strength	1.447 GPa
Transverse tensile strength	0.0517 GPa
Transverse compressive strength	0.206 GPa
In plane shear strength	0.093 GPa
Density	1605.434 kg/m^3

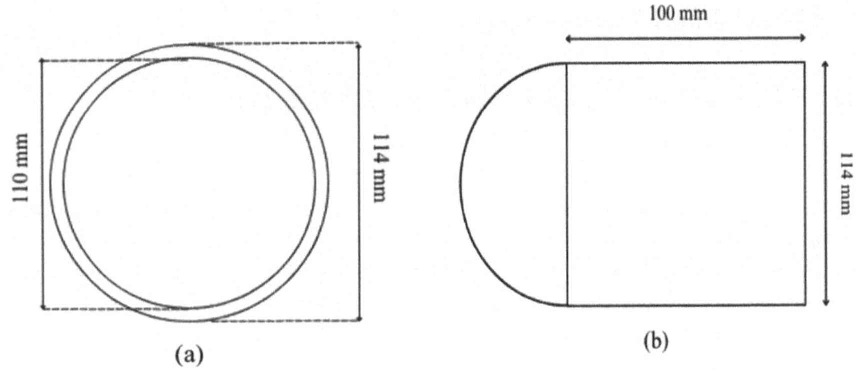

Fig. 1 **a** Top view and **b** side view of the cylindrical vessel along with dimensions

strength of 880 MPa, making them an ideal material for high-performance aerospace applications.

2.2 Finite Element Modeling and Boundary Conditions

In this study, a cylinder with an inner diameter of 110 mm, outer diameter of 114 mm, and length of the cylinder as 100 mm have been considered, as shown in Fig. 1. The finite element analysis (FEA) is performed in ANSYS workbench using ACP and static structural [5]. Finite element meshing is done in the model by using shell element meshing method because of the thin thickness and curved structure of the cylindrical vessel, as shown in Fig. 2. In the model, a 12 mm thickness of ply orientation [0°/ ± 30°/ ± 45°/90°]s and 15 mm thickness of ply orientation [0°/ ± 45°/90°]s has been considered. After that, the ply layers have been stacked together to form the overall composite combustion chamber of 8.4 cm thickness for ply orientation [0°/ ± 30°/ ± 45°/90°]s and a 7.5 cm thickness for ply orientation [0°/ ± 45°/90°]s. For this present work, a cylindrical model of thickness 1 cm has been considered only, along with a ceramic lining of inner thickness 2 cm. The cylindrical combustion chamber has been subjected to an [12] internal pressure of 50 MPa radially and have been modeled with "fixed" boundary condition on one end, as shown in Fig. 3.

2.3 Fiber Orientation for Graphite/Epoxy Composite

In the present analysis, two types of fiber layups have been considered. In the first one, a symmetric laminate structure in which the ply is arranged in [0°/ ± 45°/90°]s arrangement has been considered and in the arrangement of the second one, the plys are arranged in [0°/ ± 30°/ ± 45°/90°]s, as shown in Fig. 4.

Fig. 2 Isometric view of the finite element mesh of the cylindrical vessel

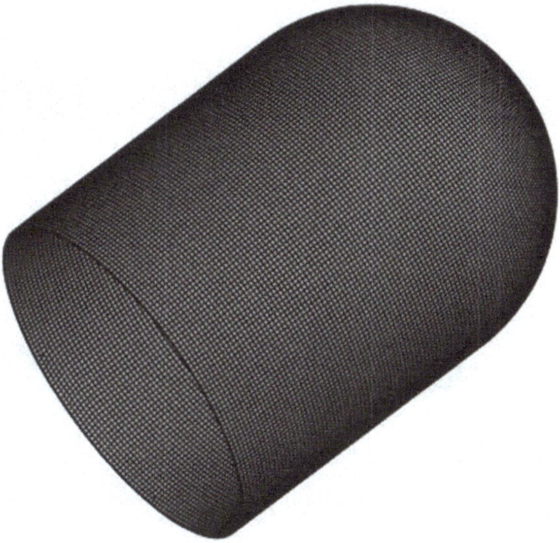

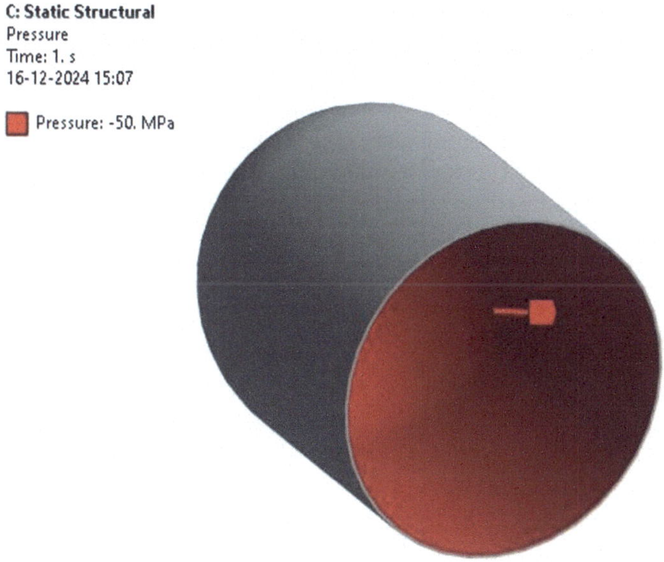

Fig. 3 Cylindrical vessel with an internal pressure of 50 MPa

2.4 Properties of Different Ply Orientations:

The [0°/ ± 45°/90°]s and [0°/ ± 30°/ ± 45°/90°]s ply orientations are the most optimized choices for solid-propellant missile combustion chambers due to their superior ability to withstand multi-directional stresses, thermal expansion, and fatigue loads

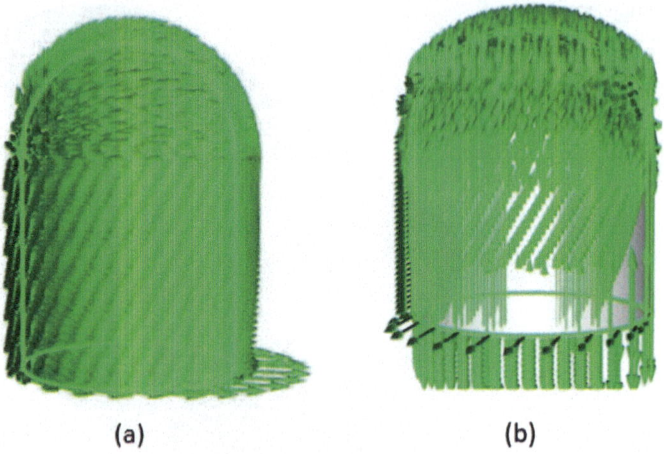

Fig. 4 **a** Fiber orientation for 0° ply and **b** Fiber orientation for 90° ply

while maintaining a lightweight structure [9]. The 0° fibers provide high axial stiffness to resist longitudinal loads generated by internal gas pressure and thrust forces, while the 90° fibers effectively counteract hoop stresses, preventing radial expansion under extreme combustion pressures. The ± 45° fibers are critical for shear load resistance, mitigating interlaminar stresses and torsional forces, which enhances the structural integrity and damage tolerance of the chamber. The addition of ± 30° fibers in the [0°/ ± 30°/ ± 45°/90°]s layup further improves shear distribution and fatigue resistance, reducing stress concentrations and extending operational life [10, 13]. These optimized orientations also help balance thermal expansion mismatches, minimizing the risk of delamination or matrix cracking due to cyclic heating and cooling. Compared to random or non-optimized layups, the [0°/ ± 45°/90°]s and [0°/ ± 30°/ ± 45°/90°]s laminates provide an ideal combination of mechanical strength, thermal stability, and manufacturability, making them the best choice for high-performance missile combustion chambers. The structural properties of the laminates at the various orientation have been determined using the classical lamination theory for each of the fiber orientation [14].

The Classical Lamination Theory (CLT) [10] is used to determine the local stresses and local strains that are acting on the different fiber orientations, respectively. The global stresses that are acting on the composite are found theoretically using the concept of thin pressure vessels [15] and from there we get that σx, σy and τxy are 137.5 MPa, 275 MPa, and 0 MPa, respectively.

$$\begin{bmatrix} \sigma_1 \\ \sigma_2 \\ \tau_{12} \end{bmatrix} = [\mathrm{T}] \begin{bmatrix} \sigma_x \\ \sigma_y \\ \tau_{xy} \end{bmatrix} \tag{2}$$

Now by using (2), the local stresses *σ1, σ2,* and *τ12,* respectively, for the different fiber orientations can be evaluated. A detailed explanation of all the terms of the Eq. (2) have been explained in the reference [10]. For the 0º fiber orientation, the local stresses are 137.5 MPa, 275 MPa, and 0 MPa, respectively. Similarly, for the 30º fiber orientation, the local stresses are 173.25 MPa, 243.37 MPa, and 59.81 MPa, respectively. For 45º fiber orientation, the local stresses are 206.25 MPa, 206.25 MPa, and 68.75 MPa, respectively, and lastly for the fiber orientation 90º, the local stresses are 275 MPa, 137.5 MPa, and 0 MPa, respectively. After that using (3) and (4), the local strains for the different fiber orientations have been determined. A detailed explanation of the equations has been given in the reference [10].

$$\begin{bmatrix} \sigma_x \\ \sigma_y \\ \sigma_{xy} \end{bmatrix} = \begin{bmatrix} \overline{Q}_{11} & \overline{Q}_{12} & \overline{Q}_{16} \\ \overline{Q}_{12} & \overline{Q}_{22} & \overline{Q}_{26} \\ \overline{Q}_{11} & \overline{Q}_{16} & \overline{Q}_{x6} \end{bmatrix} \begin{bmatrix} \varepsilon_x \\ \varepsilon_y \\ \gamma_{xy} \end{bmatrix} \tag{3}$$

$$\begin{bmatrix} \varepsilon_1 \\ \varepsilon_2 \\ \gamma_{12} \end{bmatrix} = [\mathrm{T}] \begin{bmatrix} \varepsilon_x \\ \varepsilon_y \\ \gamma_{xy} \end{bmatrix} \tag{4}$$

Solving the various local strains for each of the fiber orientation can be determined such as for 0° fiber orientation the *ε1, ε2,* and *γ12* are 0.353, 29.81, and 0 μmm/mm, respectively. Similarly, for 30° fiber orientation, the local strains are −6.828, 32.846, and 0.133 μmm/mm, respectively, for the 45° fiber orientation, the local strains are 11.257, −27.76, and 3.365 μmm/mm, respectively, and lastly for the 90° fiber orientation, the local strains that gets generated are 1.64, 14.45, and 0 μmm/mm, respectively.

2.5 *Mesh Convergence Analysis*

To assess the reliability, accuracy and robustness of the results, mesh convergence analysis has been performed. The number of elements is kept on increasing from 4,623 till the total deformation gets converged, as shown in Fig. 5 and finally the analysis is performed at 50,185 number of elements and 49,991 number of nodes. The variation of the total deformation with the number of elements is negligible. Though an almost constant deformation has been achieved at 30,000 elements, but the present analysis has been performed at 50,185 elements in order to obtain better results.

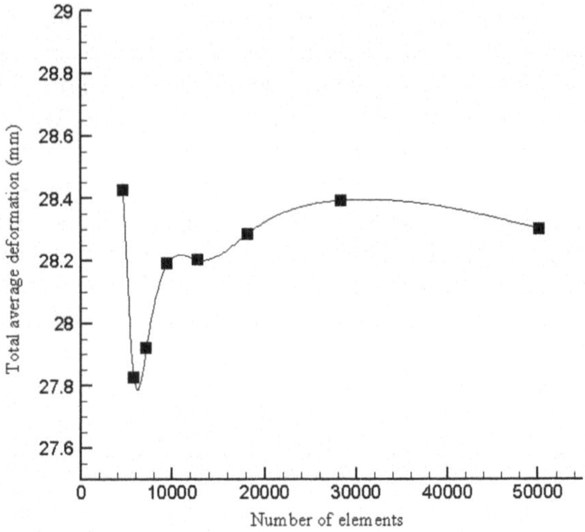

Fig. 5 Variation of total average deformation with the number of elements

2.6 *Validation of Finite Element Model*

To validate the accuracy of the FE model, structural analysis of a cylindrical vessel made of structural steel is carried out and the result is compared with theoretical calculations. In this analysis, the same amount of internal pressure of 50 MPa is applied radially in the cylindrical vessel made of structural steel with a permissible yield stress of 240 MPa. It is observed that the maximum hoop stress obtained from the theoretical results using (5) is around 2750 MPa for 1 cm thickness. Similarly, from FEA, the maximum hoop stress has been found to be around 2927.2 MPa for 1 cm thickness, as shown in Fig. 6, which yields a negligible error of approximately 6%.

$$\sigma = \frac{pd}{2t} \tag{5}$$

where $p =$ internal pressure, $d =$ diameter, and $t =$ thickness of the cylinder.

Similarly for different thicknesses, the plots were made between simulations and theoretical results and it can be seen that they were very close to each other, as shown in Fig. 7.

This analysis validates the present FE model for structural analysis of the cylindrical combustion chamber in the motor mount of the first stage of our prototype missile.

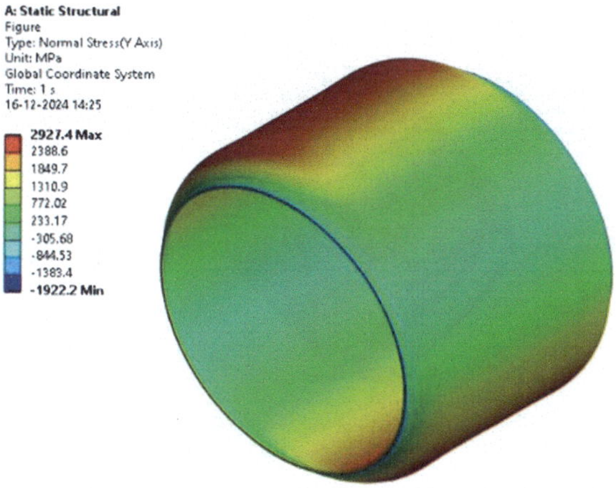

Fig. 6 Hoop stress in cylindrical vessel

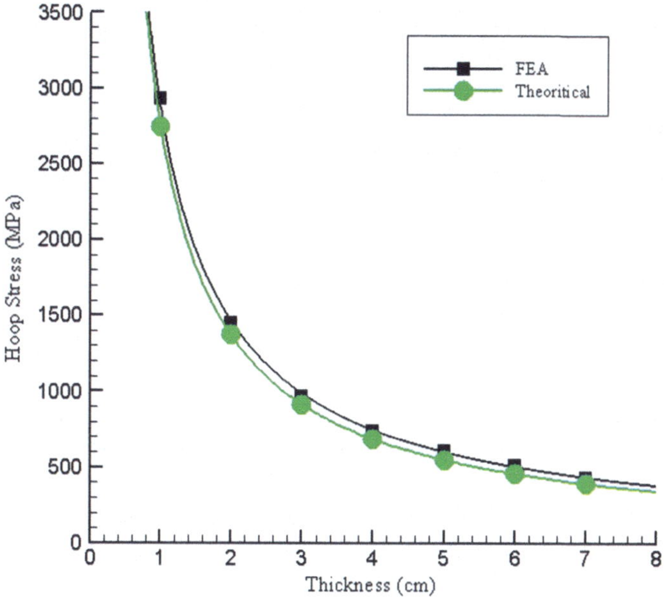

Fig. 7 Variation of hoop stress with the thickness of cylindrical vessel

3 Results and Discussion

The structural analysis performed on the cylindrical vessel made of GF epoxy composite and titanium alloy discloses important insights into the mechanical properties of these materials. It sheds light on the composite's resilience and ability to withstand high structural loads in comparison with titanium alloy, which can arguably

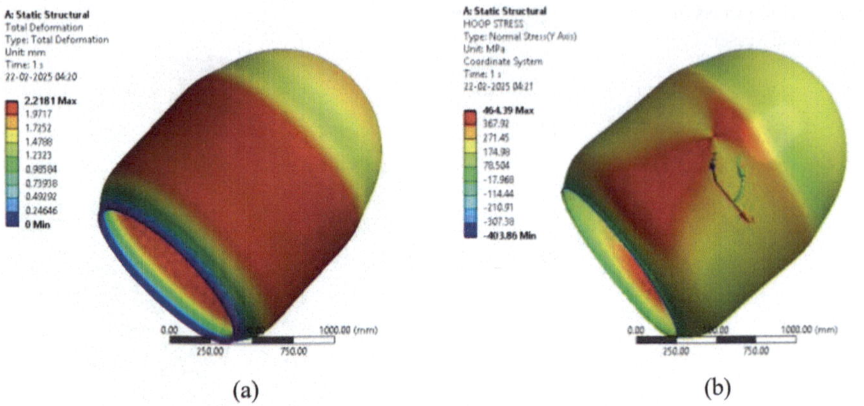

Fig. 8 **a** Total deformation, **b** Hoop stress in the cylindrical vessel made of titanium alloy

be called one of the traditional materials used in the aerospace sector [11]. The deformation and hoop stress induced in the cylindrical vessel due to internal pressure is studied to understand the structural integrity of the cylindrical vessel under varying applied structural loads.

3.1 Finite Element Analysis of the Cylindrical Vessel Made of Titanium Alloy

The total deformation shows how the cylindrical vessel changes its shape under structural load, while hoop stress analysis indicates how a material or structure is prone to behaving under structural loading. The FEA of the titanium alloy shows the total deformation and hoop stress which have occurred in the model.

From FEA, it has been observed that the titanium alloy model experiences a maximum total deformation of 2.2181 mm and an average deformation of 1.8093 mm, as shown in Fig. 8a. Also, from Fig. 8b, it is evident that the maximum hoop stress that is generated in the titanium alloy is around 464.39 MPa.

3.2 Finite Element Analysis of the Cylindrical Vessel Made of Laminate with Ply Orientation [0°/ ± 45°/90°]s

The FEA of the cylindrical vessel made of laminated composite with ply orientation [0°/ ± 45°/90°]s has been performed using the same boundary conditions as described in Sect. 2.2. The deformation occurred in the composite cylindrical vessel due to internal pressure displays impressive high specific stiffness and dimensional

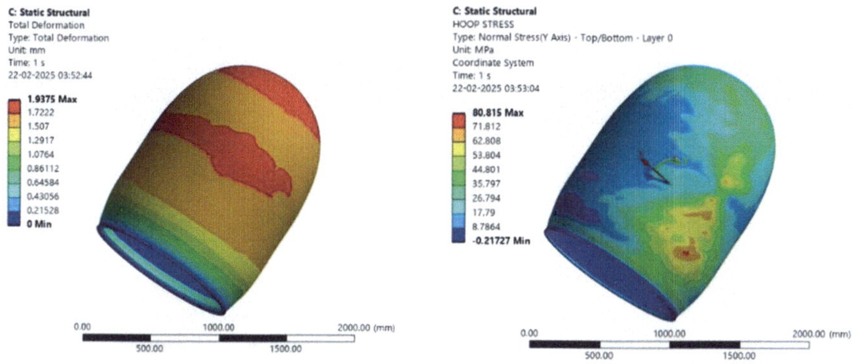

(a) (b)

Fig. 9 **a** Total deformation, **b** Hoop stress experienced in the cylindrical vessel made of laminate with ply orientation [0°/ ± 45°/90°]s

stability. It can be observed from Fig. 9a that the vessel undergoes a maximum total deformation of 1.9375 mm for the laminate orientation of [0°/ ± 45°/90°]s and it is observed from Fig. 9b that the vessel undergoes a maximum hoop stress of 80.815 MPa.

From FEA, it has been observed that the cylindrical vessel with ply orientation [0°/ ± 45°/90°]s experiences an average deformation of 1.4433 mm.

3.3 Finite Element Analysis of the Cylindrical Vessel with Laminate Made of Ply Orientation [0°/ ± 30°/ ± 45°/ 90°]s

Similarly, it can be observed from Fig. 10a and b that the vessel undergoes a maximum total deformation of 1.3945 mm for the laminate orientation of [0°/ ± 30°/ ± 45°/ 90°]s and a maximum hoop stress of 72.127 MPa.

From FEA, it has been observed that the cylindrical vessel with ply orientation [0°/ ± 30°/ ± 45°/90°]s experiences an average deformation of 1.0317 mm.

4 Comparison of the Results

The plot in Fig. 11 shows the relationship between internal pressure and total average deformation for three materials. A comparison of total deformation has been presented among the three materials: titanium alloy, graphite/epoxy composite [0°/ ± 45°/90°]s, and graphite/epoxy composite [0°/ ± 30°/ ± 45°/90°]s. This analysis

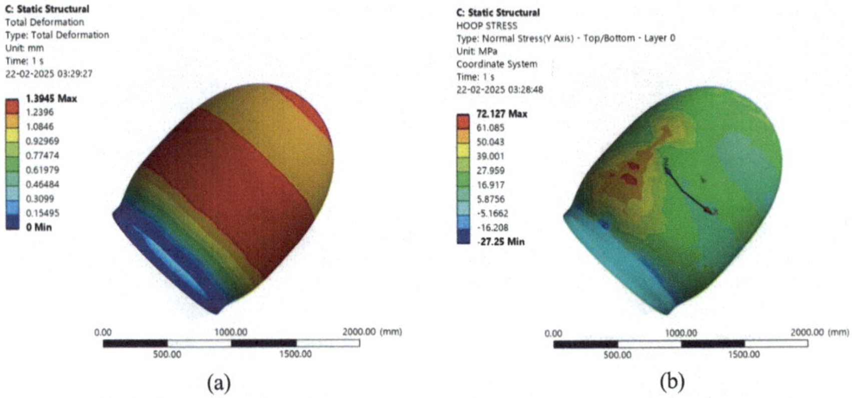

Fig. 10 **A** Total deformation, **b** Total hoop stress in the cylindrical vessel made of laminate with ply orientation [0°/ ± 30°/ ± 45°/90°]s

has been conducted to verify the structural integrity of the specified materials. This comparison highlights the influence of material composition and fiber orientation on deformation behavior under pressure.

It can be seen from Fig. 11 that as internal pressure increases, the deformation also increases for all materials, but at different rates. The titanium alloy exhibits the

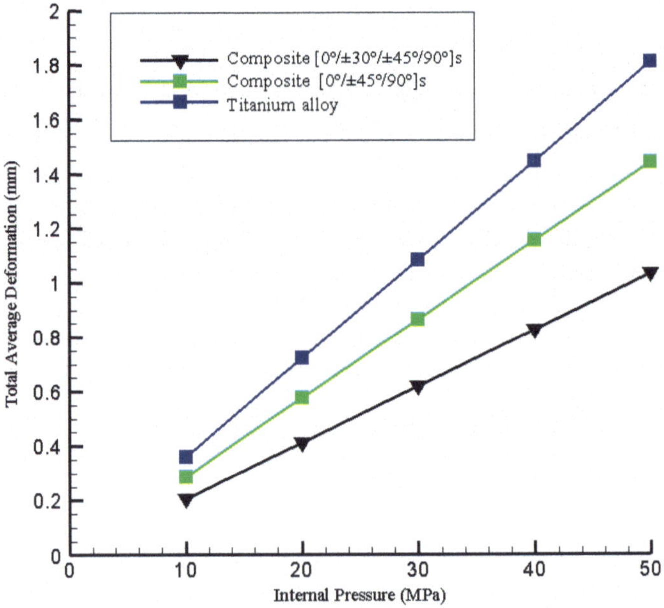

Fig. 11 Variation of total average deformation with internal pressure

highest deformation, with a steep linear increase, reaching approximately 2.21 mm at internal pressure of 50 MPa. The graphite/epoxy composite with a fiber orientation of [0°/ ± 45°/90°]s demonstrates moderate deformation, lying between the other two materials and reaching about 1.93 mm at 50 MPa. The graphite/epoxy composite with a fiber orientation of [0°/ ± 30°/ ± 45°/90°]s, shows the least deformation, indicating greater resistance to pressure-induced deformation, with deformation of approximately 1.39 mm at an internal pressure of 50 MPa. Hence, it may be concluded that the graphite/epoxy composite with a fiber orientation of [0°/ ± 30°/ ± 45°/90°]s showcases superior resistance to deformation in high-stress conditions.

The present work contributes significantly to the field of material research by investigating the application of varying ply orientation in graphite/epoxy composites for the design of machine components under structural loading. By focusing on varying fiber orientations, specifically [0°/ ± 45°/90°]s and [0°/ ± 30°/ ± 45°/90°]s, the present work highlights how these configurations influence the structural performance of missile components, namely combustion chamber, under operational conditions. Using FEA, the research develops and validates a robust model to analyze deformation and hoop stress, comparing the results with theoretical calculations to ensure accuracy. The study also provides a comparative evaluation of graphite/epoxy composites against traditional titanium alloys, demonstrating the composites' superior strength-to-weight ratio, reduced deformation, and enhanced resilience under high-stress conditions. The findings emphasize that the [0°/ ± 30°/ ± 45°/90°]s fiber orientation offers optimal performance, making it an ideal choice for missile combustion chambers, considering the mechanical load applied in the chamber. Overall, this research advances understanding of composite materials' potential in aerospace applications, offering insights into material optimization and design validation for enhanced safety and efficiency.

5 Conclusions

The present work focusses on selecting the optimum material for designing the combustion chamber in the motor mount of the first stage of our prototype missile. A comparison has been made between titanium alloy and two different types of laminated ply orientation of graphite fiber epoxy composite material and the conclusions drawn from the research are mentioned below.

- The present FE model has been validated with theoretical calculations, which yields a negligible error of approximately 6%.
- The highest average total deformation is found for the combustion chamber made of titanium alloy.
- The lowest average total deformation is found for the combustion chamber made of composite with ply orientation [0°/ ± 30°/ ± 45°/90°]s, suggesting superior resistance to deformation in high-stress condition.

- The highest total hoop stress experienced is found for the combustion chamber made of titanium alloy.
- The lowest total hoop stress experienced is found for the combustion chamber made of composite with ply orientation [0°/ ± 30°/ ± 45°/90°]s.
- The graphite/epoxy composite materials outperform the titanium alloy in terms of total average deformation and total hoop stress experienced, making them a better choice for combustion chamber applications.
- In the present work, the [0°/ ± 30°/ ± 45°/90°]s fiber orientation provides the best performance in terms of minimal average deformation and least hoop stress experienced, making them a better choice as a material for missile combustion chamber application.

References

1. Patole R, Ambhore N, Agrawal D (2023) Carbon composites in aerospace application—A comprehensive review. Mater Int 5(4)
2. Zhang J, Lin G, Vaidya U, Wang H (2023) Past, present and future prospective of global carbon fibre composite developments and applications. Compos B Eng 250:110463
3. Park S, Seo M (2011) Interface science and composites. Elsevier
4. Lukez R, Thiokol M (1987) The use of graphite/epoxy composite structures in space applications. In: small satellite conference, USA
5. Lakshmi K, Rao ARM (2013) Multi-objective optimal design of laminated composite skirt using hybrid NSGA, In: Lakshmi K, Rao ARM (eds.), Meccanica, 48(6), pp 1431–1450. Springer
6. Nath A, Anand R, Desai J, Sultan MTH, Sakthivel AR (2021) Modelling and finite element analysis of an aircraft wing using composite laminates. In: IOP conference series: materials science and engineering. 1183(1) 012006. IOP Publishing, Czech Republic
7. Jamison RD, Griffin OH Jr, Ecker JA, Skullney WE (1986) Use of graphite/epoxy composites in spacecraft structures: A case study. J Hopkins APL Tech Dig 7(3):290–294
8. Yasaka T, Onoda J (2001) Encyclopedia of physical science and technology, 3rd edn. Academic Press Inc., USA
9. Rana S, Fangueiro R (2016) Advanced composite materials for aerospace engineering: processing, properties, and applications. 1st edn. Woodhead Publishing
10. Kaw AK (2006) Mechanics of composite materials. 2nd edn. Taylor & Francis
11. Gallegos AAG, Mandal P, Gonzalez D, Zuelli N, Blackwell P (2018) Studies on Titanium alloys for aerospace application. Defects Diffus Forum 385:419–423
12. Parida SP, Jena PC (2019) Design and finite element analysis of thick-walled laminated composite pressure vessel. Int J Innov Technol Explor Eng 8(10):4389–4394
13. Bordoloi MM, Kirtania S, Banerjee S, Kashyap S (2022) Analysis of stress through the thickness of hybrid laminated nanocomposites using finite element method. J Inst Eng (India): Ser D104, 51–59
14. Baldieri F, Martelli E, Riccio A (2023) A numerical study on carbon-fibre-reinforced composite cylindrical skirts for solid propeller rockets. Polymers 15:908
15. Bhandari VB (2020) Design of machine elements, 5th edn. Mc Graw Hill, India

BoneNet: Bone Fracture Detection Using Deep Learning with Advancing Speed and Accuracy in X-Ray Analysis

Aayush Arora, Koushik Paul, Raj Saurav Parida, Ananya Bisoi, Debansh Hota, Arijit Bhattacharjee, and Abhradeep Hazra

Abstract This study reports BoneNet, an advanced deep learning framework particularly tailored to classify accurately the condition of bone with high-resolution capabilities in the assessment of bone fracture diagnosis. Bone-related pathologies critically necessitate early detection; hence, two personalized models, BoneNetV1 and BoneNetV2, were developed, engineered in such a manner that would efficiently distinguish and classify structures of bones in medical imaging. BoneNetV1 is developed using the architecture with Conv2D and LeakyReLU layers under a multiple-layer convolutional scheme, and its testing accuracy is reported to be 95.54%. BoneNetV2 combines more depth and optimized layer configurations which enhance the testing accuracy up to 99.12%. These models are highly tested by comparison with the conventional transfer learning models—VGG16, VGG19, ResNet50, DenseNet, InceptionV3, MobileNet, EfficientNet and Xception—showing that BoneNet greatly outperforms these models in metrics precision, recall, and F1 scores. The novelty of BoneNet lies in its architecture specially designed to capture intricate features of bone structure with advanced accuracy and reliability over classical approaches. Such findings put BoneNet as a powerful and reliable solution for the automated detection of bone condition, thus placing it in the potential role of aiding early and precise diagnostic practices in clinical application.

Keywords Deep learning · Convolutional architecture · Classification · Diagnostic capabilities · Transfer learning

A. Arora (✉) · A. Bisoi · D. Hota · A. Bhattacharjee
School of Computer Science Engineering, KIIT Deemed to Be University, Bhubaneswar, Odisha, India
e-mail: aayusharora0304@gmail.com

K. Paul · R. S. Parida · A. Hazra
School of Mechanical Engineering, KIIT Deemed to Be University, Bhubaneswar, Odisha, India

A. Biswas et al. (eds.), *Recent Advancements in Artificial Intelligence and Optimization*, Lecture Notes in Mechanical Engineering,
https://doi.org/10.1007/978-981-95-9268-5_11

1 Introduction

Automated analysis of bone X-rays is an emerging technology area in medical imaging with the aim of enhancing speed and accuracy in fracture diagnosis and classification. A significant health issue across the world, from osteoporosis to aging populations through sports injuries and accidental trauma, fractures account for serious morbidity, quality-of-life impairment, and, occasionally, mortality. Identification and classification of fractures as accurately and early as possible can help avoid complications, reduce the length of recovery periods, and enhance patient outcomes. Usually, conventional diagnostic techniques rely on radiologists to inspect the X-rays manually, a time-consuming process prone to subjective errors. Automated fracture detection systems have a great promise with the facts that their diagnoses are always consistent and accurate. Moreover, they usually take less time compared to standard approaches. For instance, they find great value in busy health care settings, where trusted assessments accomplished fast are indispensable. In that case, supplemental or potentially streamlined radiological assessment could help avoid misdiagnosis, reduce waiting times for patients, and facilitate faster medical workflows, which may contribute to more effective treatment and quicker recuperation of the patient [1].

Computer-aided diagnostics (CAD) systems in fracture diagnosis development are dramatically changing the field of bone imaging. It has significant advantages over traditional methods, utilizing machine learning techniques, especially deep learning models, to analyze more X-ray images for fractures more rapidly and accurately than the traditional approach. Apart from saving time, this automation helps reduce the human error associated with these processes, which can be pretty valuable, in scenarios that are poorly equipped and scant on medical aids, such as in rural India [2, 3]. CAD systems are able to identify fractures at an earlier stage than what previously used methods could, thus improving diagnostics and allowing quicker treatment. This is vital in reducing the dangers of complication due to undiagnosed fractures that may result in permanent disability or lengthen the recovery process [4]. Additionally, the system can help alleviate overworked radiologists' problems since it consistently delivers reliable results even under pressure [5].

This capability has made it possible for CAD systems to detect even the slightest fractures that in other ways might be let off by even routine manual assessments. Therefore, besides timely diagnosis, it ensures that treatment is started early, getting better prognosis among patients. Such systems are turning out to be a necessary tool in health care, especially in areas where qualified radiologists may not be readily available. We shall test many architectures of CNNs that include VGG16, VGG19, ResNet50, InceptionV3, DenseNet, MobileNet, EfficientNet, Xception as well as novel Convolutional Networks which we developed called BoneNetV1 and BoneNetV2 to differentiate and classify different types of bone fractures from images from an X-ray. We further elaborate our proprietary methodology that achieved a fantastic 99.12% (BoneNetV2) accuracy rate in our dataset of bone fracture detection. Therefore, the presented work contributes to any literature that may already exist when comparing and discussing the models' efficacy and ability to be applied in clinical practice.

This study makes a unique contribution by developing two novel convolutional neural networks: BoneNetV1 and BoneNetV2, which are specifically tailored for the task of bone fracture detection. Unlike existing models, our architectures are optimized for the reasons of achieving higher accuracy with faster detection in different clinical environments, mainly targeted at the urgent needs of under-resourced healthcare settings. This, our proprietary methodology not only offers an excellent accuracy rate of 99.12% but also sets the grounds for future innovation in automated radiological diagnostics. This research lays the grounds to further integrate deep learning solutions in routine clinical practices and thus benefit the patient in the long run.

The rest of the study is arranged as: The state-of-the-art methods developed till now are considered in Sect. 2. The network and its description that was designed is included in Sect. 3. Network performance through empirical estimation is carried out in Sect. 4. Other state of the art is performed in Sect. 5. Lastly, the article is concluded in Sect. 6 with future work directions.

2 Literature Review

The model for hand bone fractures detection in X-ray images was proposed by Mahmoud Al-Ayyoub, Ismail Hmeidi, and Haya Rababah in JMPT 4.3 pages 155–168, in 2013. The framework used advanced techniques of image-processing, including edge detection and region-based segmentation. The model applied the preprocessing steps, including noise reduction and image enhancement. These steps were taken in X-ray images to improve their performance, thus possibly increasing the identification of small fractures. The paper showed a high enough accuracy rate to gain much potential for this type of automated hand bone fracture diagnosis in clinical fields [6]. Likewise, in 2011, Hum Y. Chai et al. published a bone fracture detection method using Gray-Level Co-occurrence Matrix (GLCM) techniques in the American Journal of Applied Sciences, 8.1, page 26. The authors apply GLCM in the texture analysis, which is further aimed at capturing patterns and spatial relationships, further aiding fracture detection from X-ray images. This enhanced fracture in the proposed model, in this case, was improved by considering different texture features like contrast and homogeneity, which otherwise may not be well detected. The study demonstrated its capability regarding how good GLCM can make diagnosis more accurate by indicating possible reliability in bone clinical fracture identification [7].

In 2022, Wong, Lee, and Kim proposed an AI model with CNN to identify wrist fractures, using X-ray images. The researchers used an annotated dataset for the image classification and performed data augmentation on it. In the case of imbalanced classes, researchers used different techniques, and this model happened to yield an accuracy of 90.1% in the classification of fractures on the distal radius for people across different age groups. This study proved that CNN-based models are efficient in identifying fracture cases while decreasing human time in terms of analysis and increasing its utility even more for further improving the accuracy of diagnosis in bone imaging using AI Wong [8]. In 2023, Patel and Sharma proposed the application of the architecture called DenseNet towards developing a multi-class classification model for identifying the number of fractures of the femur, tibia, and humerus in an X-ray. The efficiency of DenseNet was combined with transfer learning, where pre-trained weights have been used to result in efficient models. The model achieved an accuracy of 92.8% on an independent test set and proved promising results for practical clinical application. Such findings show the strength of transfer learning in the detection of fractures, especially for resource-poor regions that have less availability of data Patel and Sharma [9].

Recently, a new AI algorithm has been trained by Lopez et al. in 2024 for the detection of fractures in X-rays of the foot and ankle. This study exploited 3D convolution techniques and an attention mechanism to selectively direct its attention towards specific regions with higher fracture propensity. Its overall detection accuracy was 93.5% and stands as the best into developing an early detection system for fractures in the foot. The results high-light the importance of attention-based deep learning architecture as it shows a very important ability in the accurate detection of fractures and improvement in patient outcomes in orthopedic diagnostics, according to Lopez et al [10]. In 2024, Julius Husarek and colleagues published their paper, "Artificial Intelligence in Commercial Fracture Detection Products: A Systematic Review and Meta-Analysis of Diagnostic Test Accuracy," where they presented a systematic review and meta-analysis on AI-based fracture detection products. This study assessed the diagnostic test accuracy of various commercially available AI tools applied to fracture detection: their sensitivity, specificity, and general diagnostic performance on various datasets. A meta-analysis conducted on all these studies reported a pooled estimated accuracy of 89.5%, while stating that AI-powered commercial solutions are indeed reliable in fracture detection, though there is still a need for further improvement in both clinical reliability and application. This work epitomizes the use of AI in medical imaging and the ever-increasing possibilities it provides to improve the accuracy of fracture diagnosis [11] (Table 1).

3 Methodology

In the following section, the dataset and algorithms employed for bone fracture classification from X-ray images in the bone fracture dataset have been explained. There are six sub-sect. 3.1. Dataset Used, wherein the original dataset utilized for this study has been handled, B Dataset Visualization, wherein a detailed analysis of the dataset is performed. In Sect. 3.3 Data Augmentation and Segmentation explains the data augmentation methods and preprocessing steps, and Sect. 3.4 discusses the methodologies that include Transfer Learning models. Section 3.5 introduces our proposed models BoneNetV1 and BoneNetV2. Finally, Sect. 3.6 shows the overview of software and hardware utilized for bone fracture classification.

Table 1 Literature review matrix

Paper	Merits	Limitations
Mahmoud Al-Ayyoub, et al. [6]	Used advanced image-processing techniques (edge detection, region-based segmentation) for hand bone fracture detection. High accuracy for clinical application	Limited to hand bone fractures; relies on traditional image processing rather than deep learning
Hum Chai et al. [7]	Applied Gray-Level Co-occurrence Matrix (GLCM) for texture analysis to improve fracture detection accuracy	Focused on texture features only; may miss fractures that aren't emphasized by texture analysis
Wong et al. [8]	Utilized CNNs for accurate wrist fracture detection with 90.1% accuracy; effective across age groups	Model focused only on wrist fractures; generalizability to other bones not tested
Patel and Sharma [9]	Used DenseNet with transfer learning for multi-class fracture detection (femur, tibia, humerus) achieving 92.8% accuracy	Limited to specific bone types; transfer learning results may vary with different datasets
Lopez et al. [10]	Applied attention-based deep learning for foot and ankle fractures, achieving 93.5% accuracy	Primarily focused on foot and ankle fractures; may require additional training for other fracture locations
Julius Husarek et al. [11]	Conducted meta-analysis on AI-based fracture detection products, showing 89.5% pooled accuracy for commercial tools	Suggested further improvements needed for clinical reliability and application in diverse cases

3.1 Dataset

The dataset was obtained from Kaggle and made publicly available by Darabi [12]. It contains 1,129 images of different types of bone fractures and was preprocessed to improve quality before being classified. Images in the dataset have been labeled into classes representing different fractures. A sample image from the dataset of bone fractures is represented by Fig. 1.

3.2 Dataset Visualization

This bone fracture image dataset is an image set with labels for all types of fractures across the different bones. It is a variety of cases for the classification of fracture images.

- *Images:* X-ray images of various bone fractures in JPEG or PNG format.

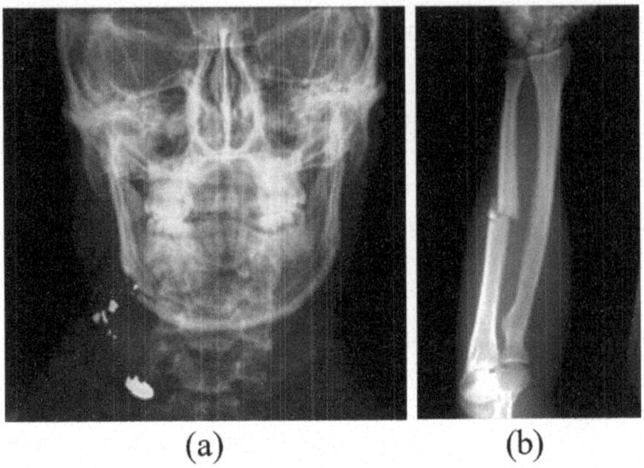

Fig. 1 X-ray images from dataset

- *Labels:* The labels for each of the images are special kind of fractures such as Avulsion fracture, Comminuted fracture, Greenstick fracture, Spiral fracture, and so on including those which ranged across different bones.

Exploratory Data Analysis (EDA) on this images only bone fracture dataset includes checking up whether images about fractures types and bones were widely disseminated. EDA with Class Frequencies and Sample Images Visualizing class frequencies along with examining sample images can help detect potential class imbalances as well as variations in image quality or orientation. The analysis assists in the effective design of preprocessing steps, such as normalization and augmentation, which is one of the most important aspects of building robust classification models. Analyzing the structural pattern of a dataset through EDA would improve model performance in accurately classifying between the types of fractures being studied and its ability to obviate the particular difficulties of the X-ray image data. Table 2 provides an overview of the distribution of images across different fracture types in the dataset (Figs. 2, 3).

3.3 Data Augmentation and Segmentation

Data augmentation fights class imbalance by augmenting the distribution of data, thus making it balanced for different classes of fractures. This has been done by producing different samples of the given images based on altered properties. In this study, it has been seen that classes, such as "Fracture Dislocation" and "Comminuted Fracture," are very high and overrepresented compared to the others, thereby causing an imbalanced dataset. This dataset is balanced by using data augmentation. Additional samples were produced by applying techniques like shearing, brightness adjustment, zoom in and out, shifting, rotation, cropping, and flipping. This makes the deep learning model stronger and unbiased. Table 3 illustrates various parameters used for data augmentation.

Table 2 Overview of the distribution of images

Fracture type (Class)	Total images
Avulsion fracture	120
Comminuted fracture	142
Fracture dislocation	163
Greenstick fracture	122
Hairline fracture	112
Impacted fracture	84
Longitudinal fracture	88
Oblique fracture	85
Pathological fracture	127
Spiral fracture	86
Total	1129

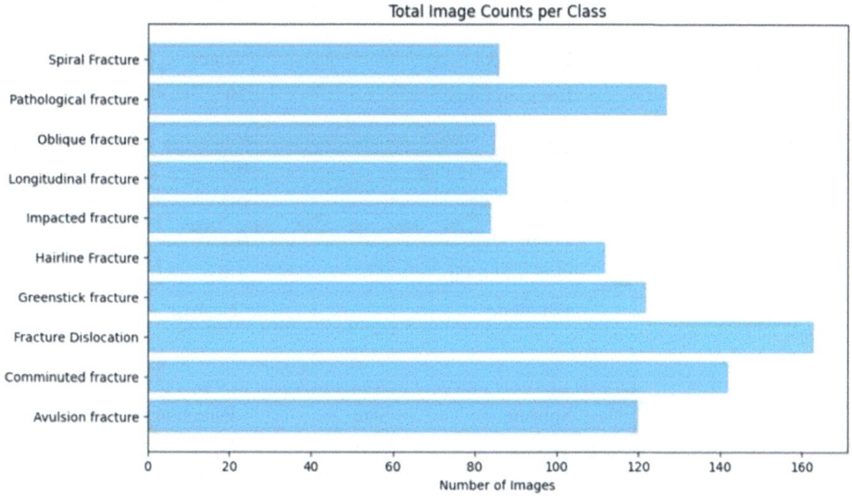

Fig. 2 Classes in bone fracture dataset

The dataset was divided into two sets: training and testing. A split of around 90–10 was used, with 90% of the data, that is, 1,017 images, assigned to training and the rest 10%, that is, 112 images, to testing. This way, a fairly sized set could be used for training the model, and a part of it would remain free to serve as unbiased testing, preventing the model from overfitting. Table 4 gives the details of Dataset for Simulation.

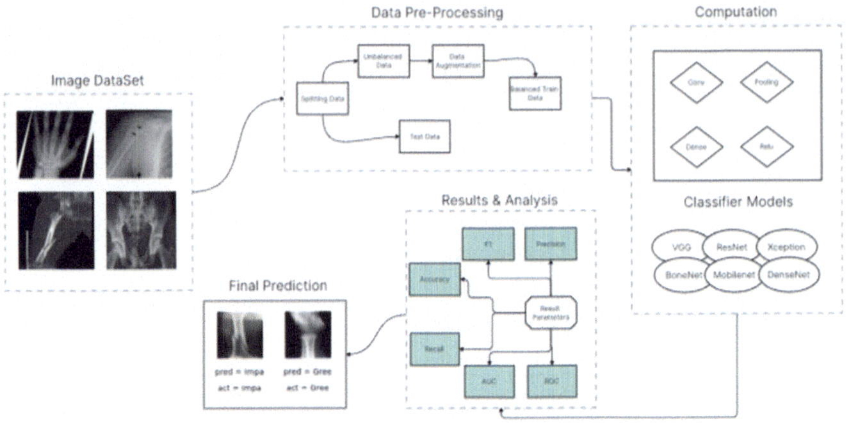

Fig. 3 Workflow for classification of bone fractures

Table 3 Parameters used for data augmentation

Parameters	Values
Rotation range	20
Width shift range	0.2
Height shift range	0.2
Shear range	0.2
Zoom range	0.2
Horizontal flip	True
Fill mode	"Nearest"

Table 4 Splitting metrics

Splitting	Size	Percentage (%)
Training	1,017	90
Testing	112	10
Total	1,129	100

3.4 Transfer Learning Models Used

In this study, we examined a few of the advanced transfer learning models for bone analysis: VGG16, VGG19, ResNet50, InceptionV3, DenseNet, MobileNet, EfficientNet, and Xception. These models are picked since they have different architectures and feature extraction capabilities. This exploration will enable readers to peek deeply into each of the models' contributions in terms of accuracy in classification, computational efficiency, and their adaptability towards medical imaging requirements, thereby helping this study form a solid basis for future applications in the diagnosis of bones.

Vgg16

VGG16 is a very deep but highly simple architecture convolutional neural network that uses sequential layers of 3 × 3 convolutions using fixed configuration to catch fine-grained features across various scales in images. Its design is also uniform since it allows uniform extraction of features and spatial hierarchies which can be useful for high-resolution tasks such as bone analysis [13]. However, VGG16 has a very high memory and computational cost. Therefore, it will be less ideal in real-time applications as compared to lighter models like MobileNetG19.

Vgg19

VGG19 is an extension of the previously mentioned VGG16 model with three more convolutional layers, which increases its depth and ability to capture more precise patterns in images. It is beneficial in detailed medical image tasks because the deeper architecture allows VGG19 to have improved feature extraction, particularly for complex textures [13]. However, unlike VGG16, VGG19 is computationally intensive and uses more processing capacity, which puts quite a limit on its efficiency in low resource environments.

ResNet50

The residual connections in the architecture of ResNet50 obviate the problem of vanishing gradients in very deep networks. In essence, such an architecture enables the effective training of deeper models without any reduction in accuracy, making ResNet50 one of the ideal architectures for the task of medical imaging, which aims to derive deeper features [14]. In essence, based on residual connections, ResNet50 does not lose any information by bypassing some layers, and hence it is one of the most versatile models applicable to bone analysis and other medical applications.

InceptionV3

InceptionV3 employs a quite novel module design that involves multiple filter sizes which are applied in parallel. This approach helps capture diverse spatial features in images more effectively. The multi-scale approach can take in fine and broad patterns that enhance the utility of analyzing the wide variety of textures present in bone images [15]. InceptionV3's design also optimizes computation with an efficiency suited for balance-biased tasks between accuracy and resource usage.

DenseNet

In DenseNet, dense connections include all the previous layers to every later layer, allowing a better gradient flow and feature reuse. This makes it easier for DenseNet to get high accuracy with fewer parameters and is especially helpful for applications dealing with detailed and complex textures such as bone structures [16]. Further increasing its ability to capture details at the cost of some computational efficiency, this is also possible through an efficient use of parameters in DenseNet.

MobileNet

MobileNet is computationally inefficient and instead optimized for mobile and resource-constrained environments. Using depthwise separable convolutions, it significantly reduces the cost of computation without losing a strong feature-extracting capability, making it applicable to real-time medical imaging applications [17]. Considering its lightweight architecture, MobileNet is well-suited for light computation environments with limited resources, such as portable diagnostic devices.

EfficientNet

The scaling method is used in EfficientNet so that each dimension of it balances depth, width, and resolution so that the model reaches at high accuracy with optimized computational efficiency [18]. Therefore, this model would be well-suited for the applications requiring precision as well as efficiency to arrive at an important conclusion together. Because of uniform scaling ability, the model of EfficientNet gives robust performance across the difference datasets and resolutions, and due to it, it is highly versatile for tasks requiring detailed analysis, such as bone imaging.

Xception

The Xception uses depthwise separable convolutions. This cuts down the computationally expensive nature and, in turn, improves feature extraction capabilities. With a high level of accuracy, the model may represent the representation of fine-grained patterns and, consequently, be quite effective for detailed medical images [19]. Xception differs in design, as it may achieve comparable or even higher accuracies than much larger models without, in any way, imposing the same computational burdens. It is a very strong candidate for high-resolution tasks like bone analysis. Table 5 represents the metrics for transfer learning models.

Table 5 Transfer learning models

Model	Parameter (Mi)	Layers
VGG19	24.554	19
VGG16	16.895	16
DenseNet121	82.26	121
Inception V3	57.96	48
ResNet50	32.06	50
EfficientNet	176.04	82
Mobile Net	7.584	28
Xception	293.5	71

3.5 *Proposed Networks*

In this paper, we introduce two customized models, namely, BoneNetV1 and BoneNetV2, especially for bone analysis. The model of BoneNetV1 plays a foundation role since it applies a well-balanced architecture with optimum layers for an efficient feature extraction on images of bone. Building from that, BoneNetV2 added layers and further defined specific configurations in the model to capture more detailed structures of bone and improve overall accuracy and performance. Both the models have been fine-tuned to the special needs of bone analysis with an objective of improved performance than the traditional transfer learning models.

BoneNetV1

A design for bone imaging analysis aims to optimize the same by detailed capturing of features in medical images through a sequential convolutional architecture. It includes an input layer for 256 × 256x3 images and is comprised of several layers of convolution, starting from 32 up to 256 filters. Each individual layer applied a 3 × 3 kernel with strides of 2; such settings maximize spatial reduction while obtaining critical features. Batch normalization with a momentum set at 0.9 is applied to all the layers without any variation to stabilize and speed up the training. Finally, leaky ReLU activations are used with a value of 0.2 to provide better gradient flow and handle sparse features in the bone structures. After every convolutional layer, dropout layers are set with a rate of 0.2 that ensures that the model does not overfit enabling the network to generalize well without too many epochs. Figure 4 gives an overview to the network architecture of the BoneNetV1 network.

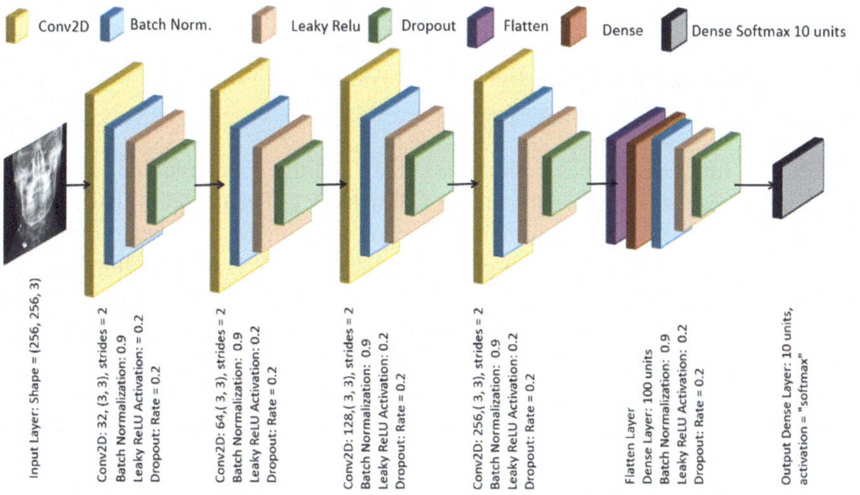

Fig. 4 Network architecture of the BoneNetV1

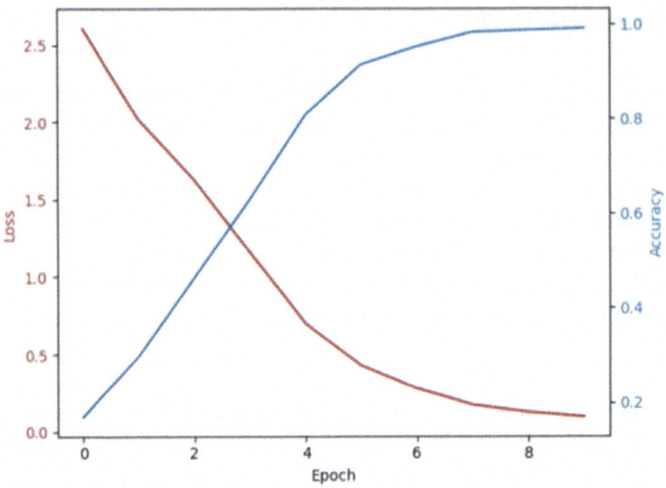

Fig. 5 Loss and accuracy of BoneNetV1

Its novelty is that it hits such a balance between depth and regularization that higher accuracy with lesser computational cost is achieved. BoneNetV1 flattened the output after the last convolutional layer, fed it into a dense layer of 100 units for further feature combination before the final softmax output layer, which sets an impressive accuracy of 95.54% in just 10 epochs. This makes this model ideal for bones analysis tasks with a strong need for high accuracy, but not requiring too long training times: simple yet effective convolutional blocks with normalization and dropout. Such architecture is a solid foundation upon which further developments are made to capture even more subtle features in BoneNetV2. Figure 5 shows the graphics of accuracy and loss of BoneNetV1 over number of epochs.

BoneNetV2

BoneNetV2 further enhances both the accuracy and the computational efficiency, which advances the overall architecture of BoneNetV1. The model initiates with a Conv2D layer with 60 filters of kernel size 3 × 3 along with ReLU activation, pushing the network towards learning richer patterns in bone images. A MaxPooling2D layer then downsamples the spatial dimensions and detects major features. Next follows the second Conv2D layer with 120 filters, which increases the depth in the feature extraction process, followed by another MaxPooling2D, reducing the dimensionality of the model very efficiently with the saving of important features in the process. Figure 6 gives an overview to the network architecture of the BoneNetV2 network.

This is where BoneNetV2 boasts its new signature: reduced complexity and greater accuracy at 99.12% in just 20 epochs. After the last pooling layer, there is a Flatten layer. It converts spatial dimensions into fully connected layers. For these fully connected layers, it includes one dense layer with 40 units, along with a ReLU activation function. Finally, to do multi-class classification, the model makes

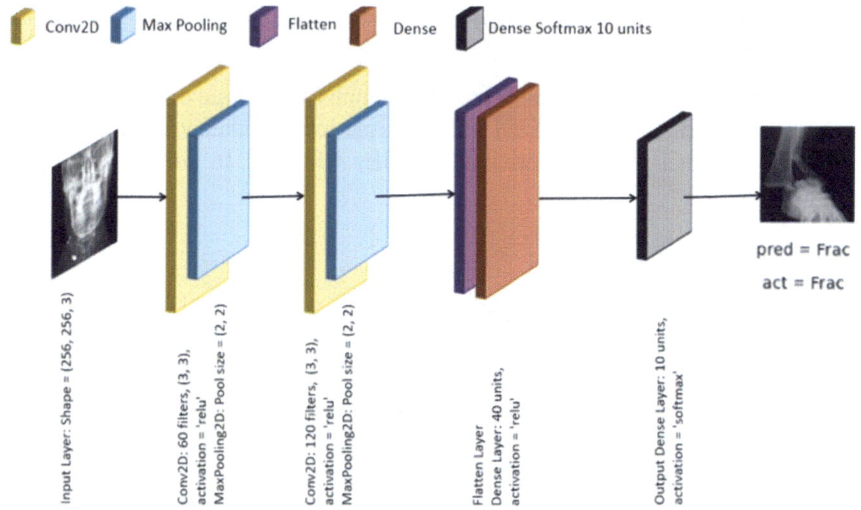

Fig. 6 Network architecture of the BoneNetV2

use of an output layer which has a softmax activation function. BoneNetV2 balances the number of parameters and network depth to yield a lightweight but powerful solution that is well-suited for applications of bone analysis which require a much higher accuracy with efficient processing. Figure 7 shows the graphics of accuracy and loss of BoneNetV2 over number of epochs.

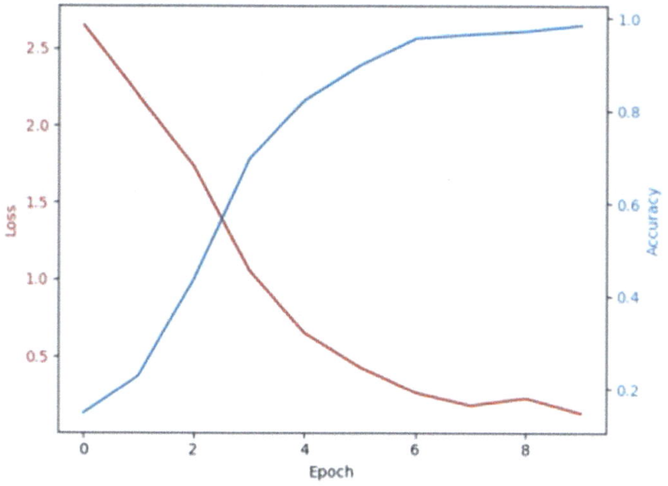

Fig. 7 Loss & accuracy of BoneNetV2

3.6 Deployment Environment

All of the deep learning deployments were accomplished and run on Python 3 using Keras and TensorFlow were utilized with a Kaggle Jupyter notebook. The hardware workstation is Intel i711th generation 2.80 GHz CPU and 16 GB RAM running Windows 11.

4 Implementation and Results

In this section, we present the implementation details and experimental results obtained using BoneNetV1 and BoneNetV2 as well as other transfer learning models, designed to analyze bones based on deep learning techniques. The implementation stage varies with data preprocessing architecture of the model and relevant training parameters chosen to improve the performance of the models. These are evaluated for accuracy and computational efficiency, where the different models have proved to classify bone images up to a great extent. In this section, we will fully describe the experimental settings and performance metrics while pointing out improvements that we have achieved using BoneNetV2 as compared to the other models. Table 6 shows performance metrics of different models.

It is observed that the transfer learning models are performing differently in classifying bone images. The precision for VGG16 is moderate with a recall of 0.86, leading to an F1 score of 0.84, while the precision was 84.82% during the test. However, VGG19 performs poor with its precision and recall values set at 0.36 and 0.35, respectively, while the F1 score is 0.35, resulting in just a 35.18% accuracy. ResNet50 is comparable with VGG16. ResNet50 also exhibits a precision of 0.85, a recall of 0.84, and an F1 score of 0.84 and a testing accuracy of 83.93%. MobileNet has very poor performance—0.42 for precision, 0.39 for recall, and a value of 0.36

Table 6 Performance metrics for models

Model name	Precision	Recall	F1 Score	TestingAcc. (%)	Complexity
VGG16	0.86	0.85	0.84	84.82	111
VGG19	0.36	0.35	0.35	35.18	332
ResNet50	0.85	0.84	0.84	83.93	183
Mobile Net	0.42	0.39	0.36	36.61	224
Inception Net	0.33	0.30	0.26	29.46	385
Efficient Net	0.56	0.47	0.46	48.21	213
Xception	0.96	0.95	0.95	94.64	469
Dense Net	0.91	0.91	0.91	91.07	159
Bone Net V1	0.96	0.96	0.96	95.54	362
Bone Net V2	0.99	0.99	0.99	99.12	134

for F1 score, with an accuracy of 36.61%. InceptionNet is not so good at this task—it gives 0.33 for precision, a value of 0.30 for recall, and a value of 0.26 for F1 score, and finally, it shows only 29.46% accuracy.

For EfficientNet, the model resulted in a moderate improvement with a precision of 0.56, recall of 0.47 and an F1 score of 0.46. In terms of testing accuracy, it reached 48.21%. Xception brought about excellent results. It obtained high precision with a score of 0.96, recall with a score of 0.95 and an F1 score of 0.95. Testing accuracy gained 94.64%. DenseNet also presented good results. All the scores attained 0.91, which ultimately obtained 91.07% accuracy. The proposed BoneNet models significantly outperform other transfer learning models. BoneNetV1 achieves a precision, recall, and F1 score of 0.96. The accuracy achieved on the testing set is 95.54%. BoneNetV2 further improves and shows superior results. For example, its precision, recall, and F1 score reach 0.99, and the achieved testing accuracy is 99.12%. Such superior results are enough to establish the superiority of BoneNetV2 as an extremely efficient model for bone image classification with robust generalization and accuracy in key metrics (Fig. 8).

The study tests the bone fracture classification with various classes of bone fracture conditions, BoneNetV2 demonstrated its advanced possibility of discrimination between the highly complex bone structures. Results with perfect precision, recall, and 99% overall accuracy thereby underlie the novelty and effectiveness of BoneNetV2 to achieve high performance levels with minimal error rates. As demonstrated in the Fig. 9, BoneNetV2 classifies bone images accurately, which, in turn, gives evidence about its ability to generalize to different types of bones that can be found within a given dataset. Based on the results of this study, the capability and reliability of this model are ensured, and therefore, it can be certified as the ideal choice for bone image analysis tasks. The architecture of BoneNetV2 collaborates efficient

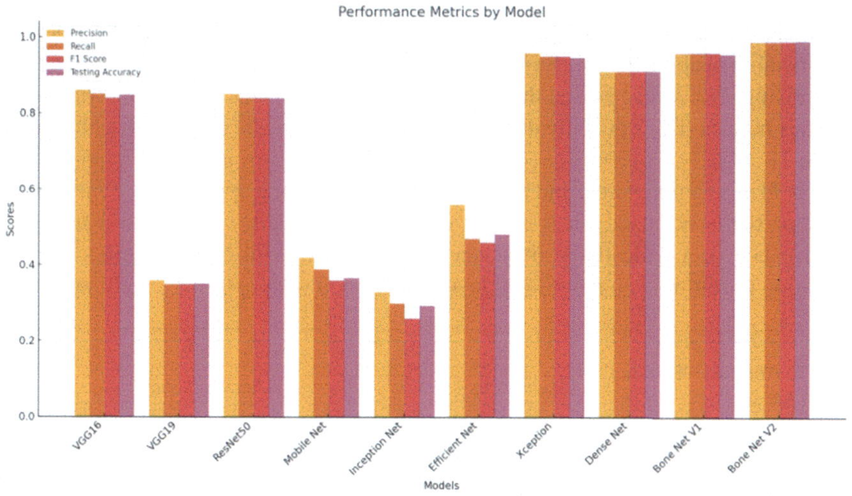

Fig. 8 Performance of all models

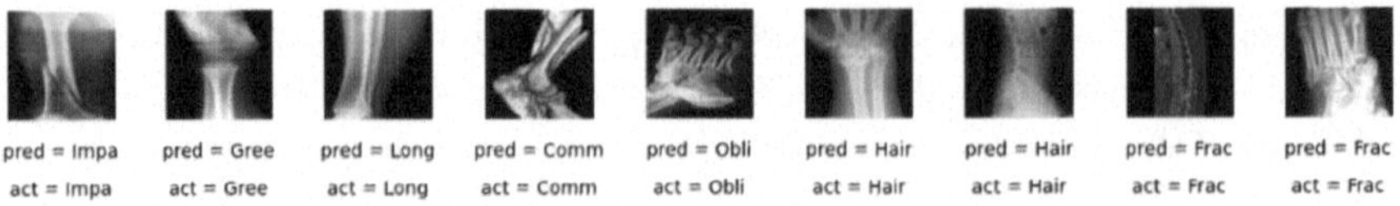

Fig. 9 Prediction of actual labels

feature extraction with deeper layers that certainly makes it go beyond the current transfer learning models to achieve better performance in classification. Capable of performance enhancement in classification, the new approach brings the model into a very premier position for real-world applications concerning clinical diagnostics, especially earlier and more accurate bone-related condition identification, which may be helpful in the improvement of patient outcomes.

5 Comparison with Other State of the Art

BoneNet performed better than the model proposed by Farid et al. [20], by achieving 92.3% of accuracy. In that result, it has displayed that BoneNet generalizes very well with an accuracy of 99.12%. This defines the effectiveness and efficiency of BoneNet compared to the proposed model, especially in handling the complexity of the fracture diagnosis. BoneNet clearly outperforms the study at 93.7% accuracy by Breu et al. [21], using AI-assisted diagnosis. BoneNet generalizes well to achieve a high accuracy of 99%. This proves that BoneNet is as effective and efficient in comparison to the introduced model, particularly on the improvement of accuracy as well as the speed of fracture diagnosis.

BoneNet fully beats it by comparison to the 2024 experiment on Ankle fracture classification by Olczak et al. [22], which had an accuracy for an ankle fracture classification of about 91.2%. BoneNet generalizes very well that its accuracy reaches a surprising 99%. This proves the effectiveness and efficiency that BoneNet holds over this new model, especially its advancement in orthopedic diagnostic technology (Table 7).

BoneNet is an innovative and highly adaptive classifying approach of bone conditions, particularly based on specially designed architecture that would improve performance in a wide range of scenarios associated with bone imaging. Implementation of multiple convolutional layers combined with optimized techniques for

Table 7 Analysis with other state of the arts

Techniques	Accuracy (%)
Farid et al. [20]	92.3
Breu et al. [21]	93.7
Olczak et al. [22]	91.2
Our work	99.12

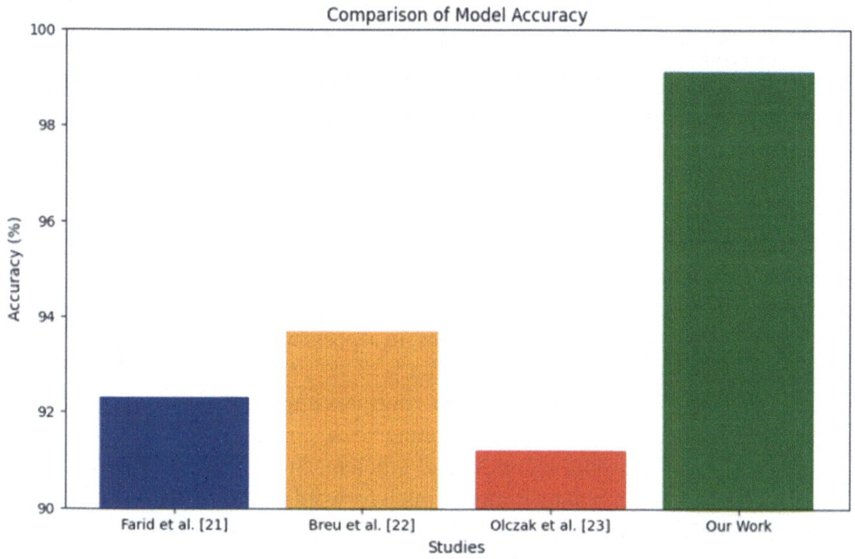

Fig. 10 Performance analysis with other state of arts

feature extraction signifies the ability of BoneNet to prove robust feature representation learning even under varied imaging conditions caused by differences in equipment or resolution, and even patient demographics. This flexibility is critical in medical imaging, in which such variability can substantially change the quality and interpretability of images. Another good evidence is sustained high testing accuracy, whereby the model sustains performance at 95.54% for BoneNetV1 and 99.12% for BoneNetV2. Such architectural choices go further to enhance the robustness of BoneNet in handling domain shifts, hence ensuring robust performance in the challenging clinical environment. The comparison with other up-to-date models further emphasizes its high adaptability and efficiency, as deduced from Fig. 10.

Overall, BoneNet works much better not only in its precision but also in its adaptability and sturdiness in a large number of conditions, and thus it has a great potential as a tough contender for bone fracture classification in comparison with current models. This sturdiness and flexibility will guarantee that it can meet the very diverse requirements of medical professionals for the treatment and diagnosis of fracture classifications.

6 Conclusion and Future Work

- This study introduces BoneNet—a deep learning based framework for improving bone fracture classification from medical images.

- In essence, this paper aimed at achieving high classification accuracies through optimization of the computational resources and minimizing the training time. Excellent performance was demonstrated by both BoneNetV1 and BoneNetV2, with testing accuracies of 95.54% and 99.12%, respectively.
- The models outperformed existing transfer learning models, such as VGG16, ResNet50, and other models (mentioned in Sect. 3.4) with lower accuracies and precision metrics.
- This places BoneNet as most effective solution for bone health classification. Through the use of the total appraisal metric which comprised all precisions, recalls, and F1 scores, it is found that BoneNet performs best towards generalization on different kinds of bone conditions, in aspects which in clinical applications they may rely on.
- Future work on BoneNet will include solving class imbalances in medical datasets. Future work may also include integrating these models with real-time X-ray machines to make results seamless and efficient. Strong results from the automated condition detection by BoneNet put it forward as an excellent tool for early diagnosis and intervention in bone health management.
- This work belongs to the rapidly growing topic of applications of deep learning in medical image and shall serve as a basis for further research in related fields such as bone fracture localization and multi-condition classification.

References

1. Sabino J, Raphael L (2019) Detecting fractures in radiographs with convolutional neural networks. https://cs229.stanford.edu/proj2019aut/data/assignment_308832_raw/26613897.pdf
2. Pujitha B, Raga Sravya K, Krishnasai N, Aparna C (2024) Detection of bone fracture using deep learning. In: 2024 International conference on emerging systems and intelligent computing (ESIC), Bhubaneswar, India. pp 703–708, https://doi.org/10.1109/ESIC60604.2024.10481535
3. Jung J, Dai J, Liu B, Wu Q (2024) Artificial intelligence in fracture detection with different image modalities and data types: A systematic review and meta-analysis. Artificial intelligence in fracture detection with different image modalities and data types: A systematic review and meta-analysis. PLOS Digital Health 3(1):e0000438. https://doi.org/10.1371/journal.pdig.0000438
4. Tanzi L, Vezzetti E, Moreno R, Moos S (2020) X-Ray bone fracture classification using deep learning: a baseline for designing a reliable approach. Appl Sci 10:1507. https://doi.org/10.3390/app10041507
5. Artificial intelligence in fracture detection with different image modalities and data types: A systematic review and meta-analysis Jung J, Dai J, Liu B, Wu Q (2024) Artificial intelligence in fracture detection with different image modalities and data types: A systematic review and meta-analysis. PLOS Digital Health 3(1): e0000438. https://doi.org/10.1371/journal.pdig.0000438
6. Al-Ayyoub M, Hmeidi I, Rababah H (2013) Detecting Hand Bone Fractures in X- Ray Images. J Multimed Process Technol (JMPT). 4:155–168
7. Muchtar M, Simanjuntak S, Rahmat R, Mawengkang H, Zarlis M, Sitompul O, Winanto I, Andayani U, Syahputra M, Siregar I, Nasution T (2018) Identification tibia and fibula bone

fracture location using scanline algorithm. J Phys: Conf Ser 978:012043. https://doi.org/10.1088/1742-6596/978/1/012043
8. Wong T, Lee H, Kim D (2022) AI-based model for wrist fracture detection using convolutional neural networks. J Med Imaging 15(3):125–135
9. Patel R, Sharma V (2023) Multi-class fracture classification using DenseNet and transfer learning for femur, tibia, and humerus X-ray images. Int J Med Imaging 17(2):89–98
10. Lopez M, Garcia J, Rodriguez A (2024) Attention-based deep learning for foot and ankle fracture detection in radiographic imaging. Orthopedic AI J 22(1):15–28
11. Husarek J, Hess S, Razaeian S, Ruder TD, Sehmisch S, Müller M, Liodakis E (2024) Artificial intelligence in commercial fracture detection products: a systematic review and meta-analysis of diagnostic test accuracy. Sci Rep. 14. https://doi.org/10.1038/s41598-024-73058-8
12. Darabi PK Bone Break Classification Image Dataset. Available: https://www.kaggle.com/datasets/pkdarabi/bone-break-classification-image-dataset
13. Simonyan K, Zisserman A (2014) Very deep convolutional networks for large-scale image recognition. *arXiv preprint* arXiv:1409.1556
14. He K, Zhang X, Ren S, Sun J (2016) Deep residual learning for image recognition. In: Proceedings of the IEEE conference on computer vision and pattern recognition (pp 770–778)
15. Szegedy C, Vanhoucke V, Ioffe S, Shlens J, Wojna Z (2016) Rethinking the inception architecture for computer vision. In: Proceedings of the IEEE conference on computer vision and pattern recognition (pp 2818–2826)
16. Huang G, Liu Z, Van Der Maaten L, Weinberger KQ (2017) Densely connected convolutional networks. In: Proceedings of the IEEE conference on computer vision and pattern recognition (pp 4700–4708)
17. Howard AG (2017) Mobilenets: Efficient convolutional neural networks for mobile vision applications. *arXiv preprint* arXiv:1704.04861
18. Tan M, Le Q (2019) Efficientnet: Rethinking model scaling for convolutional neural networks. In: International conference on machine learning (pp. 6105–6114). PMLR
19. Chollet F (2017) Xception: Deep learning with depthwise separable convolutions. In Proceedings of the IEEE conference on computer vision and pattern recognition pp 1251–1258
20. Farid Amirouche, Aashik Mathew Prosper, Majd Mzeihem et al (2024) Enhancing pediatric distal radius fracture detection: optimizing YOLOv8 with advanced ai and machine learning techniques. PREPRINT (Version 1) available at Research Square [https://doi.org/10.21203/rs.3.rs-5306607/v1]
21. Breu R, Avelar C, Bertalan Z, Grillari J, Redl H, Ljuhar R, Hausner T (2024) Artificial intelligence in traumatology: a comparative study between conventional and AI-aided diagnostic performance in distal radius fractures. Bone & Joint Research 13(10):588–595
22. Olczak J, Prijs J, IJpma F, Wallin F, Akbarian E, Doornberg J, Gordon M (2024) External validation of an artificial intelligence multi-label deep learning model capable of ankle fracture classification. BMC Musculoskelet Disord, 25(1), 788

Solar Resource Prediction Using Data-Driven Meta Models

Kalpaayan Bhattacharjee, Prayag Raj Chanda, and Agnimitra Biswas

Abstract In this work, four different machine learning models are developed for the predictions of ambient and solar data using long range (10 years) but discrete datasets in the fitting of the models. These datasets are taken from the location of this study, i.e., Silchar, Assam, India. Machine learning-based interpolation methods such as kriging interpolation, polynomial regression interpolation, nearest neighbor interpolation, and piecewise cubic Hermite interpolation techniques are used for developing the models to compare between the actual and predicted values. MATLAB code is written to generate the comparable graphs between the predicted results and interpolated results, i.e., fitted curves with actual values. The results indicate that the median errors in predictions by the models are in the range of 5.19–11.38%, with the least error exhibited by the Kriging interpolation method. Thus, this study demonstrates that kriging interpolation method can be suitably applied for the ambient and solar resource predictions for tropical conditions with discrete datasets.

Keywords Meta model · Solar prediction · Optimization · Interpolation

1 Introduction

In today's world, where climate change is a big problem, countries and communities are becoming more interested in saving energy and taking care of the environment [1]. This interest is giving countries and communities an opportunity to upgrade their energy systems and move away from on fossil fuels. Instead, they're looking at using alternative cleaner and more environmentally friendly sources of energy. Solar energy is a limitless and eco-friendly, affordable power source available abundantly in many places around the world. The energy potential of solar resources in India due to its geographic location and climate [2]. The solar energy potential in India is estimated to be around 5,000 trillion kilowatt-hours per year, which is significantly higher than

K. Bhattacharjee · P. R. Chanda (✉) · A. Biswas
Department of Mechanical Engineering, National Institute of Technology Silchar, Silchar, India
e-mail: mailmeprayag@gmail.com

A. Biswas et al. (eds.), *Recent Advancements in Artificial Intelligence and Optimization*,
Lecture Notes in Mechanical Engineering,
https://doi.org/10.1007/978-981-95-9268-5_12

its current energy consumption (1,844 trillion kilowatt-hours) [3]. Machine learning (ML) methods have a vital role in the field of solar energy [4]. Their application in this context involves leveraging advanced algorithms and data-driven approaches to extract insights, make predictions, and enhance decision-making related to solar power [5].

1.1 Solar Resources and Its Variation with Weather Parameters

Solar energy, the emitted by the sun, is an abundant renewable energy source. However, the amount of solar energy that reaches the Earth's surface varies depending on several weather parameters. Understanding this variability is crucial for maximizing the efficiency of solar power generation systems. The primary solar resource is solar radiation, measured in watts per square meter (W/m2) [6]. It consists of various wavelengths of electromagnetic radiation, with the intensity of each wavelength affecting solar cell efficiency. The amount of solar radiation reaching a specific location depends on Clouds are the most significant factor affecting solar radiation [7]. Thick cloud cover can drastically reduce the amount of sunlight reaching the panels, leading to a sharp drop in power generation [8]. While solar radiation provides the energy for electricity generation, solar panel efficiency decreases with rising temperatures [9]. Higher temperatures lead to increased internal resistance within the panels, reducing their ability to convert sunlight into electricity [10]. While not a direct factor in radiation intensity, high humidity can lead to condensation on solar panels, reducing their efficiency by reflecting sunlight [11]. Understanding how weather parameters affect solar resource availability is crucial for predicting solar power output based on weather forecasts allows for better integration with the electricity grid and energy management strategies.

1.2 Importance of Solar Resource Prediction

Solar energy is rapidly becoming a cornerstone of the change towards a clean and green energy future. However, unlike traditional fossil fuels, solar power is inherently variable due to its dependence on weather patterns. This variability poses a significant challenge for grid integration and maximizing the efficiency of solar power plants. Solar resource prediction emerges as a critical tool for navigating these challenges [12]. The success of a large-scale shift towards solar energy hinges on maintaining a stable and reliable electricity grid [13]. Solar resource prediction provides crucial insights into future power generation [14].

1.2.1 Optimizing Solar Power Plant Operations

Solar resource prediction empowers solar power plant operators to make data-driven decisions that enhance efficiency and profitability [15]. This includes maintenance scheduling and power trading strategies. Maintenance scheduling means predicting periods of low solar irradiance allows for preventive maintenance tasks to be scheduled during these times, minimizing downtime during peak production periods [11]. On the other hand, power trading strategies refers to an understanding anticipated solar power output, plant operators which can participate in energy markets more effectively [16]. They can strategically buy or sell electricity based on predicted production levels and market demand.

1.2.2 Fostering a Sustainable Energy Market

The increasing penetration of solar power into the grid necessitates a more dynamic and flexible energy market. Accurate solar resource prediction contributes to improved market efficiency by providing greater certainty about future solar generation, price fluctuations in the energy market can be minimized, fostering a more stable and predictable trading environment and enhanced investment opportunities [17]. It also gives enhanced investment opportunities by providing reliable solar resource prediction data which reassures investors about the potential returns and operational feasibility of solar projects, attracting further investment in the renewable energy sector [18, 19].

2 Meta Models for Solar Resources Prediction

This section explains about different methods that has been implemented in the whole study [9]. In ordinary kriging, the estimated value at an unobserved location is given by a linear combination of the observed values at nearby locations, with weights determined by the spatial correlation structure of the data. The formula for ordinary kriging estimation at a location, $\hat{Z}^{\circ}$ is as follows [5]:

$$\hat{Z} = \sum_{i=1}^{n} \lambda_i Z_i \tag{1}$$

where Z° is the estimated value at the unobserved location. Z the observed value at the i-th location. λi is the kriging weight assigned to the i-th observation. The kriging weights, λi, are determined based on the spatial correlation structure of the data and are usually calculated using the kriging system of equations:

$$\sum_{j=1}^{n} \lambda j \gamma(\text{hij}) + \lambda^{\circ} = \gamma(\text{h}^{\circ}) \tag{2}$$

where $\gamma(hij)$ is the semi variance between locations i and j separated by a vector hij. λ° is a Lagrange multiplier to ensure that the sum of the weights is equal to 1. $\gamma(h^{\circ})$ is the semi variance between the unobserved location and each of the observed locations. Solving this system of equations yields the kriging weights, which are then used in the first formula to estimate the value at the unobserved location. It's important to note that the actual formulas may vary depending on the specific type of kriging being used (ordinary, simple, universal) and the semi variogram model employed to describe the spatial correlation. The semi variogram is a key component of kriging and describes how the variance of the data changes with distance and direction.

2.1 *Regression Interpolation*

Regression interpolation uses algorithms establish a relationship between dependent variable (also known as the target variable) and one or more independent variables (also known as features or predictors). In practice, regression learners are trained on historical data where both the input features and the corresponding target values are known. During the training process, the algorithm learns the underlying patterns and relationships in the data, which it then uses to make predictions on new, unseen data.

Polynomial regression is form of regression analysis in which the relationship between the independent variable x and the dependent variable y is modeled as an n-degree polynomial function. The formula for polynomial regression can be expressed as follows:

$$y = b_0 + b_1 x + b_2 x^2 + \ldots\ldots\ldots\ldots + b_n x^n + \varepsilon \tag{3}$$

where y is the dependent variable (the variable we are trying to predict). b_0, b_1, b_2, b_n are the coefficients of the polynomial terms. x is the independent variable (the variable we are using to make predictions). n is the degree of the polynomial. ε is the error term, representing the difference between the predicted and the actual values.

2.2 *Nearest Neighbor Interpolation*

Nearest neighbor interpolation is a straightforward and computationally efficient method for estimating unknown values in a range of a sets of known data points. This technique is particularly useful in applications where simplicity and speed are prioritized over smoothness and accuracy of the interpolated values. Given a set of known data points (x_i, y_i) for $i = 1, 2, \ldots\ldots n$, where x_i is the independent variables and y_i are the corresponding dependent variables, the goal is to estimate the value y_q at a query point x_q. Nearest neighbor interpolation can be mathematically described as

$$y_q = y_k \tag{4}$$

where y_k is the value of the nearest known data point x_k to the query point x_q.

$$k = \arg\ min_i \left|x_q - x_i\right| \tag{5}$$

Here, arg min_i represents the argument (or index) of the minimum value. It identifies the index i that minimizes the subsequent expression. $|x_q\text{-}x_i|$ represents the absolute difference between the query point x_q and each of the known data points x_i . Nearest neighbor interpolation is ideal for applications requiring rapid computation and exact replication of original data values, albeit at the expense of producing a piecewise constant interpolated function.

2.3 Piecewise Cubic Hermite Interpolation

Piecewise Cubic Hermite Interpolation (PCHIP) is a method of interpolation that constructs a smooth curve through a set of known data points. Unlike simple linear interpolation, PCHIP uses cubic polynomials to achieve a more accurate and visually appealing approximation of the data. The main advantage of PCHIP over other cubic interpolation methods (like cubic spline interpolation) is its ability to preserve the shape and monotonicity of the data. This makes PCHIP particularly useful in applications where it is crucial to avoid overshooting and to maintain the general trend of the data, such as in scientific data visualization, computer graphics, and data fitting in various engineering fields. Given a set of data points $(x_{i,}y_i)$ for i = 1,2,…,n PCHIP constructs a piecewise cubic polynomial for each interval $[x_i+x_{i+1}]$ that not only interpolates the data points but also ensures that the first derivatives of the polynomials are continuous and shape-preserving.

The PCHIP method constructs a cubic polynomial $P_i(x)$ for each interval $[x_i+ x_{i+1}]$

$$P_i(x) = a_i(x - x_i)^3 + b_i(x - x_i)^2 + c_i(x - x_i)^1 + d_i \tag{6}$$

where x_i is the independent variable values (e.g., time, position). y_i are the dependent variable values (e.g., temperature, solar irradiance) corresponding to each x_i. The coefficients a_i, b_i, c_i, and d_i are determined to ensure that the polynomial fits the data points and preserves the desired properties (smoothness, monotonicity). Here, a_i & b_i ensure the polynomial fits the interval and maintains smoothness and monotonicity.c_i ensures the slope at the data point. d_i ensures the polynomial passes through the known data point.

2.4 Errors Obtained

The accuracy of kriging depends on various factors, including the characteristics of the spatial data, the choice of the semi variogram model, and the spatial correlation structure. Assessing the error in kriging predictions is typically done through cross-validation or other validation techniques. One common measure to assess the accuracy of predictions is the Root Mean Squared Error (RMSE) [20]. The RMSE is calculated as follows:

$$\text{RMSE} = \sqrt{\frac{1}{n}\sum_{i=1}^{n}(z_i - zo_i)^2} \tag{7}$$

where n are the number of observations, Zi are the observed value at location i, zo_i is predicted (kriged) values at location i. RMSE provides a measure of the average error between the observed values and the kriging predictions. A lower RMSE indicates better predictive performance. Another commonly used measure is the Mean Absolute Percentage Error (MAPE), which is the average difference between observed and predicted values:

$$\text{MAPE} = \left|\frac{1}{n}\sum\frac{z_i - zo_i}{z_i}\right| \tag{8}$$

2.5 Ambient and Solar Data

For the purpose of this study, various parameters like solar irradiance, clearness ratio and relative humidity for the co-ordinates of NIT Silchar has been considered. The following data has been taken from NASA LARC website spanning a period of 10 years, from 2012 to 2022 [1, 12] (Table 1).

3 Results and Discussion

In this section, the results obtained for all four meta models have been represented graphically. The factors which have been investigated for solar irradiance. Here we have taken ten random dates in calendar year. After the data is obtained from different prediction models, the predicted data is validated with actual data that has been obtained from NASA LARC website. Based on the four meta models and their model equations explained in the previous chapter, in this chapter the predictions by these models are compared with their actual values. MATLAB code is written

Table 1 Various weather parameters used for this study

Month	Relative humidity (2012–2022) (%)	Clearness index (2012–2022)	Solar irradiance (2012–2022) (kW-hr/m^2/day)
January	74.85	0.58	4.61
February	65.48	0.58	5.27
March	60.44	0.57	5.54
April	73.93	0.49	4.93
May	82.89	0.43	4.44
June	88.49	0.39	4.01
July	88.16	0.42	4.31
August	88.11	0.44	4.32
September	87.72	0.47	4.53
October	84.06	0.52	4.8
November	80.44	0.62	5.19
December	79.78	0.58	4.59

to generate the comparable graphs between the predicted results and interpolated results, i.e., fitted curves using actual values.

3.1 Predicted and Interpolated Data

In Fig. 1, various interpolated and predicted values of various parameters such as solar irradiance, relative humidity and clearness are shown using krigging interpolation. Here blue curve indicates interpolated values and red dot on the blue curve indicates the predicted values.

In Fig. 2, the red curve indicates the actual direct normal irradiance that has been obtained from NASA LARC website, whereas the blue curve indicates the predicted value for some particular dates using kriging interpolation.

In Fig. 3, interpolated curve and predicted values predicted values of various parameters such as solar irradiance, relative humidity and clearness are shown. Here blue curve indicates interpolated values and red dot on the blue curve indicates the predicted values and the values are obtained using nearest neighbor interpolation method.

In Fig. 4, the red curve indicates the actual direct normal irradiance that has been obtained from NASA LARC website, whereas the blue curve indicates the predicted value for some particular dates using nearest neighbor interpolation method.

In Fig. 5, blue dots indicates then actual direct normal irradiance that has been obtained from NASA LARC website, whereas the blue curve indicates the predicted value for some particular dates using regression interpolation.

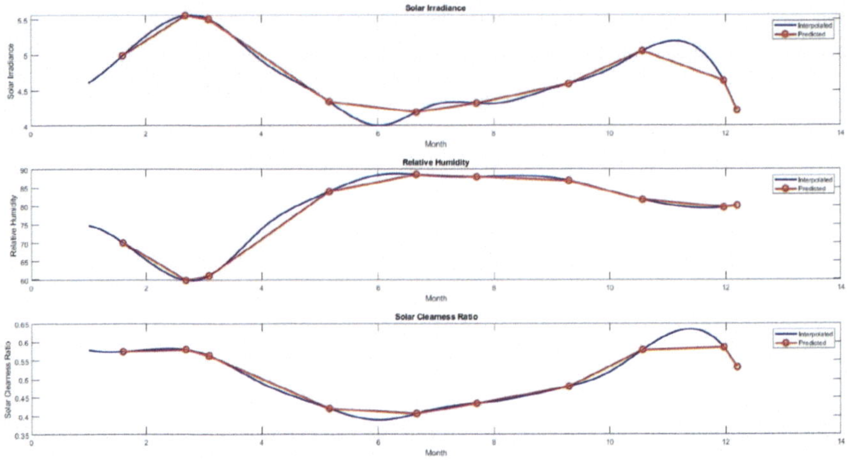

Fig. 1 Interpolated and predicted values of different parameters using kriging interpolation method

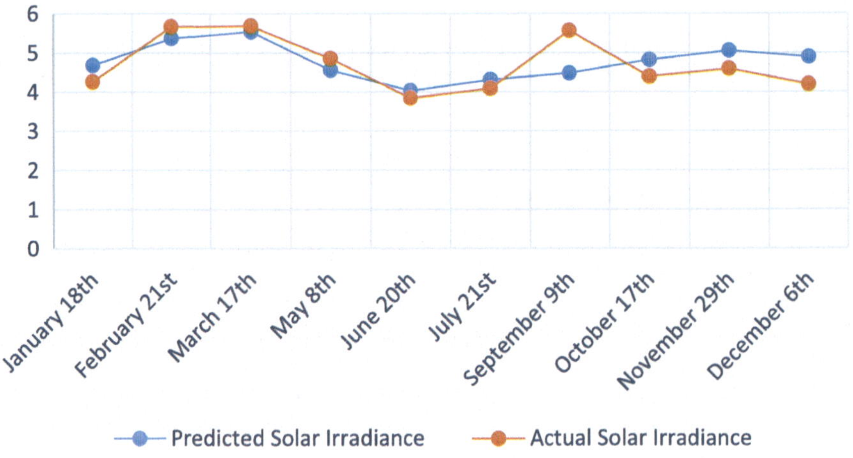

Fig. 2 Comparison of predicted and actual direct solar irradiance values using kriging interpolation

In Fig. 6, interpolated curve and predicted values of various parameters as solar irradiance, relative humidity and clearness ratio are shown and the values are obtained using piecewise cubic Hermite interpolation method.

In Fig. 7, the yellow curve indicates the actual direct normal irradiance that has been obtained from NASA LARC website, whereas the blue curve indicates the predicted value for some particular dates using piecewise herminte interpolation.

In Fig. 8, the values of actual direct solar irradiance obtained from NASA LARC, predicted values by different interpolation methods like kriging, regression, nearest neighbor, and PCHIP plotted on the graph.

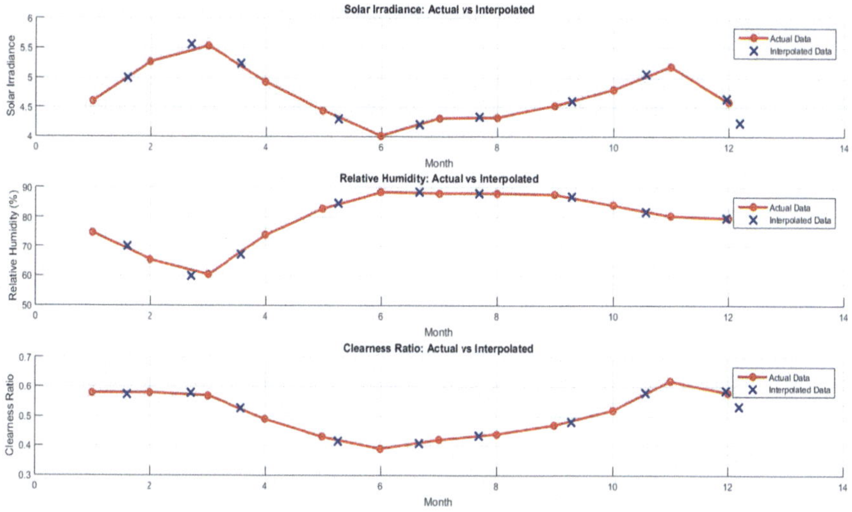

Fig. 3 Interpolated and predicted values of different parameters using nearest neighbor interpolation

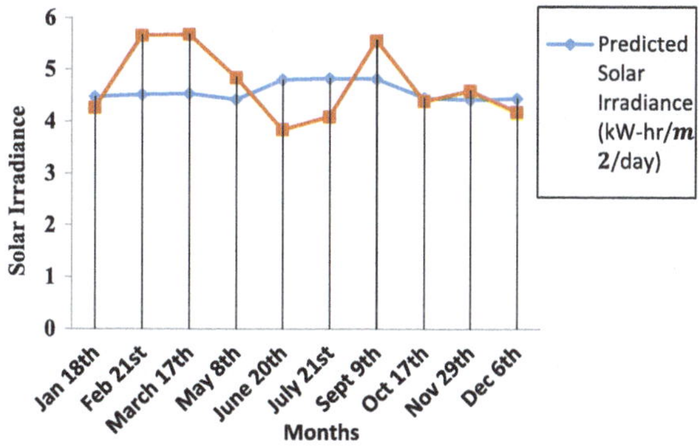

Fig. 4 Comparison between actual and predicted values using nearest neighbor interpolation

3.2 *Prediction Analysis*

For prediction of solar irradiance, ten random days of the year 2021 using different interpolation methods. The table is further divided in four parts where first column contains ten random dates for the year 2021, seconds column contains the predicted solar irradiance, the third columns contains the actual direct solar irradiance that has been obtained from NASA LARC website and the fourth column contains the percentage of error (Table 2).

After cross-verification of different interpolated data for eight random dates in a year it is seen in the table that every interpolation technique is subjected to some errors. The error caused can be validated with real world data that can be

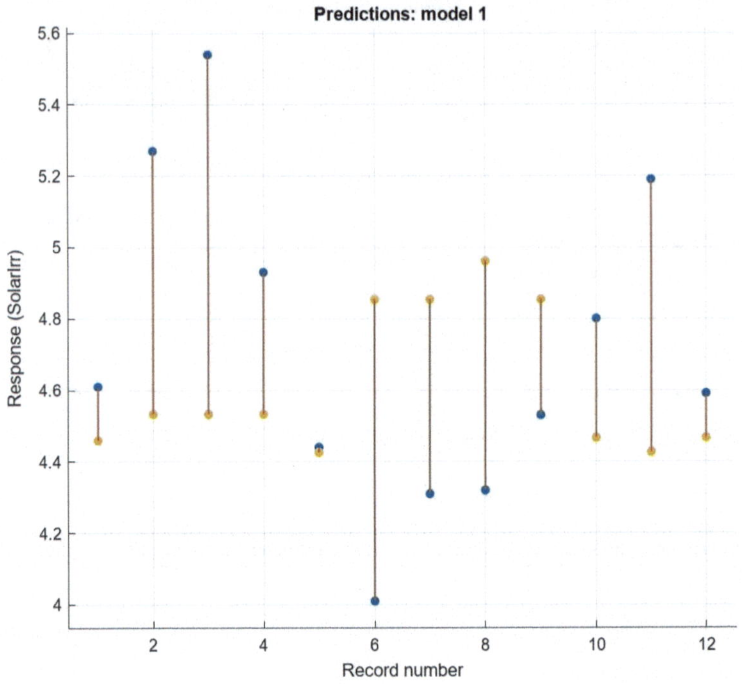

Fig. 5 Comparison of predicted and actual direct solar irradiance values using regression interpolation

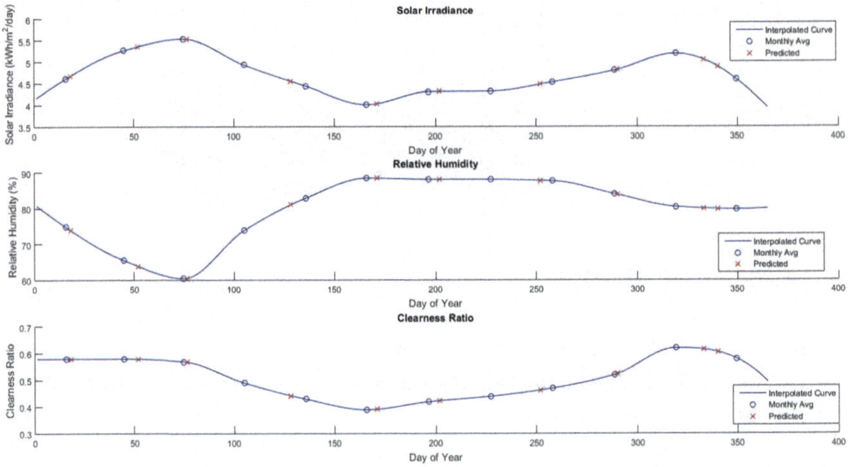

Fig. 6 Interpolated and predicted values of different parameters using piecewise cubic Hermite interpolation

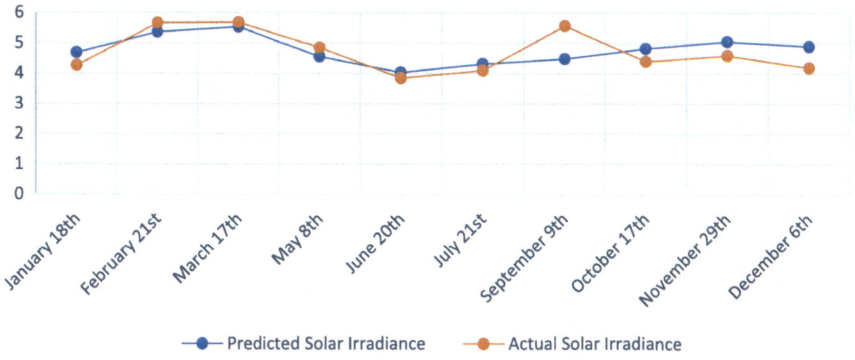

Fig. 7 Comparison between actual and predicted values using piecewise herminte interpolation

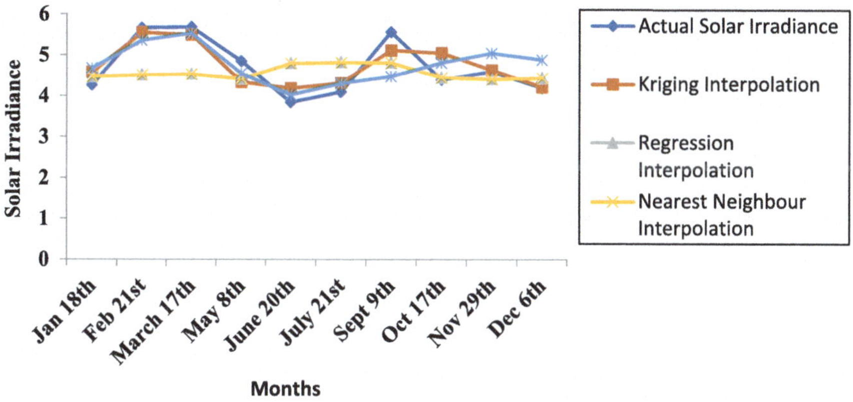

Fig. 8 Comparison between actual and predicted values of different interpolation methods

accessed from NASA LARC website. Errors are measured using RMSE method. After validating a comparative study has been carried out (Table 3).

It is observed that the interpolation methods implemented methods implemented for the study yields an error percentage ranging for 0.348–0.706 for Root Mean Square Error and 5.19%–11.38% for Mean Absolute Percentage Error. The kriging interpolation method has the least error, whereas regression interpolation method has the most.

Table 2 Actual and various predicted data using different interpolation methods

Day of the Year	Actual Solar Irradiance (kW-hr/m^2/day)	Predicted Solar Irradiance using krigging interpolation (kW-hr/m^2/day)	Predicted Solar Irradiance using nearest neighbor interpolation (kW-hr/m^2/day)	Predicted Solar Irradiance using regression interpolation (kW-hr/m^2/day)	Predicted Solar Irradiance using PCHIP (kW-hr/m^2/day)
January 18th	4.27	4.59	4.48	4.48	4.6816
February 21st	5.67	5.5589	4.52	4.52	5.3613
March 17th	5.69	5.5079	4.78	4.54	5.5370
May 8th	4.86	4.3477	4.43	4.43	4.5536
June 20th	3.85	4.197	4.81	4.81	4.0311
July 21st	4.1	4.3253	4.83	4.83	4.3127
September 9th	5.57	5.1264	4.82	4.82	4.4794
October 17th	4.4	5.0569	4.46	4.46	4.8169
November 29th	4.6	4.6414	4.42	4.42	5.0516
December 6th	4.2	4.2227	4.45	4.45	4.8901

Table 3 Comparison of errors of different interpolation techniques

SL. No	Method	Root mean square error	Mean absolute error (%)
1	Kriging interpolation	0.348	5.19
2	Regression interpolation	0.706	11.38
3	Nearest neighbor interpolation	0.670	9.13
4	Piecewise cubic hermite interpolation	0.500	7.42

4 Conclusion

This study gives a summary of all the outcomes obtained from the experiment. Here, comparison of different outputs has been obtained by using different machine learning-based techniques.

- Four different machine learning models are developed for predictions of ambient and solar data using long range (10 years) but discrete datasets in model fitting.

- In this study, out of all machine learning methods investigated it is found that kriging interpolation generated the most accurate predicted data when validated with actual data. The root mean square error for kriging interpolation technique is found to be 0.348.
- Novelty of this study is that is gives prediction from long range discrete datasets.

As the world is slowly transitioning from conventional energy sources to renewable energy sources, use of machine learning-based meta models play a very vital role in it. Some of the major scopes of machine learning-based meta models are meta models can be used to predict feasibility of solar energy for any geographical location and it can be used to efficiently measure the collector performance by reducing the experimental and computational costs. The primary limitation of this experiment is that its accuracy is heavily dependent on the size of the dataset. Larger datasets tend to produce more accurate results, whereas smaller datasets are associated with relatively higher errors.

References

1. power.larc.nasa.gov
2. Chanda PR, Podder B, Biswas A, Sengupta AR (2023) Advancements in solar assisted drying technologies: A comprehensive review post 2017. J Stored Prod Res 104:102190
3. A report on "Load Generation Balance Report (2014-15)", Ministry of power, Central Electricity Authority of India, Govt. of India. [3] A report on Economic Survey of India, 2014–15
4. Mandal P, Madhira STS, Haque AU, Meng J, Pineda RL (2012) Forecasting power output of solar Photovoltaic system using wavelet transforman artificial intelligence techniques. Procedia Comput Sci. 12, 332–337
5. Ogliari E, Grimaccia F, Leva S, Mussetta M (2013) Hybrid predictive models for accurate forecasting in PV systems. Energies 6:1918–1929
6. Borkakoti S, Das B, Gupta A (2024) Environmental and economic assessment of single-slope solar greenhouse dryer for ginger and turmeric drying in north-eastern region of India. J Stored Prod Res 106:102299
7. Aristotelis Vartholomaios (2019) A machine learning approach to modelling solar irradiation of urban and terrain 3D models computers, environment and urban systems. 78, 101387
8. Gabriel Narvaez, Luis Felipe Giraldo, Michael Bressan (2021) Andres pantoja machine learning for site-adaptation and solar radiation forecasting. 167 pp 333–342
9. Zhe Song, Sunliang Cao, Hongxing Yang (2024) An interpretable framework for modeling global solar radiation using tree-based ensemble machine learning and Shapley additive explanations methods. 364. 123238
10. Ranjith Kumara K, Surya Kalavathib M (2016) Artificial intelligence based forecast models for predicting solar power generation PMME
11. Fawaz Alassery, Ahmed Alzahrani, Asif Irshad Khan, Kashif Irshad, Saiful Islam (2022) An artificial intelligence-based solar radiation prophesy model for green energy utilization in energy management system. 52, Part B, 102060
12. Salma Zaim, Mohamed El (2023) Using artificial intelligence for global solar radiation modeling from meteorological variables. 215, 118904
13. Cyril Voyant, Gilles Notton, Soteris Kalogirou, Marie-Laure Nivet al (2017) Christophe Paoli, Fabrice Motte, Alexis Fouilloy, Machine learning methods for solar radiation forecasting: A review. 105, Pages 569–582

14. Alejandro Barredo Arrieta, Natalia Díaz-Rodríguez, Javier Del Ser (2020) Explainable Artificial Intelligence (XAI): Concepts, taxonomies, opportunities and challenges toward responsible AI. 58, Pages 82–115
15. Ashkan Entezari, Alireza Aslani (2023) Artificial intelligence and machine learning in energy systems: A bibliographic perspective. Energy Strat Rev 45(8):101017
16. Zaid Allal, Hassan N, Noura, Khaled Chahine (2024) Machine learning algorithms for solar irradiance prediction: a recent comparative study. 7, 100453
17. Neha Sehrawat, Sahil Vashisht, Amritpal Singh (2023) Solar irradiance forecasting models using machine learning techniques and digital twin: A case study with comparison. Int J Intell Netw. 4, Pages 90–102
18. Edgar Dario Obando, Sandra Caravajal, Jaior Pineda (2019) Solar radiation prediction using machine learning techniques: a review. IEEE Latin America Transactions 17(04):684–697
19. Blessing O Abisoye, Yanxia Sun, Zenghui Wang (2024) Machine learning forecasting model for solar energy radiation. Int J Comput Theory Eng, 16, 2
20. Chanda PR, Biswas A (2024) Carving a niche in building energy using modern vernacular house with passive wall materials-A multi-criteria decision-making framework. Energy 312:1335

Future Directions of Electric Discharge Drilling on Composites: A Review

Amit, Basanta Kumar Bhuyan, and Anil Kumar Dahiya

Abstract Electrical Discharge Drilling (EDD) is an efficient shaping technique that makes a variety of holes in hard materials that are challenging to drill. The current endeavor is to find the different factors to enhance the material removal rate (MRR) and decrease the tool wear rate (TWR), save the drilling time and energy, and reduce the holes to expand for the EDD process. A common non-contacting hybrid machining technique for electrically conductive materials is EDD, which finds application in nuclear power plants, automobiles, aircraft, and other construction sectors. It generates extremely high temperatures that vaporize the work material on the workpiece's surface using a high discharge voltage supply. Composite materials are now widely employed in sophisticated engineering processes. Owing to their anisotropic and non-homogeneous character, machining them presents significant challenges for these industries. The aim of this article is to focus the inquiry on the fundamental physical notion of the EDD process and analyze the experimental work of earlier researchers in tabular form. This research also reports on how input parameters affect machining characteristics and illustrates the EDD process variables using a block diagram. The scopes of the resources and research potential have been analytically represented using a pie chart based on the critical examination of the literature.

Keywords Composite materials · MRR · electric discharge drilling (EDD) · dielectric fluid · TWR

Amit · B. K. Bhuyan (✉)
Department of Mechanical Engineering, SET, MRIIRS, Faridabad, Haryana, India
e-mail: bkbhuyan.set@mriu.edu.in

A. K. Dahiya
Department of Mechanical Engineering, MAIT, Delhi, India

A. Biswas et al. (eds.), *Recent Advancements in Artificial Intelligence and Optimization*,
Lecture Notes in Mechanical Engineering,
https://doi.org/10.1007/978-981-95-9268-5_13

1 Introduction

Electrical Discharge Drilling (EDD) is a competent material removal process that uses discharge energy to melt and evaporate material form the workpiece surface. It shares the same fundamental principles as Electric Discharge Machining (EDM), with the key difference being the use of a rotating tool in EDD [1]. Electrically conductive material of any hardness may be machined using EDD, making it a valuable development in machining technology. EDD works by preventing direct contact between the tool electrode (cathode) and the workpiece (anode). Rather, a narrow space filled with dielectric fluid divides them. This spacing, sometimes called the inter-electrode gap (IEG), usually falls between 0.01 and 0.5 mm [2]. The tool electrode and the workpiece material are submerged in a dielectric fluid which clean the machining area, removing debris and also serves as a conduit for electrical discharge [3]. To form the plasma channel a high-frequency DC voltage of 20–120 V is used. Additionally, electrons which migrate from the tool electrode to the workpiece join stray electrons in the gap and increase the temperatures and pressures by dissolving the dielectric. Extreme heat and pressure in the plasma channel cause local melting and evaporation of the tool electrode and workpiece material [4]. Regular electrical discharges (sparks) between the tool and workpiece surfaces cause material removal [5]. EDD technology was initially developed for the fabrication of complex dies. Over time, it has evolved into several advanced forms such as dry EDD, hybrid EDD, and Vibration-Assisted EDD to enhance material removal rate [6]. The die-sinking electro-discharge machining is used drill holes in conductive materials [7]. The input design parameters are optimized to improve the MRR using Taguchi technique in circular cutting on steel with wire EDM [8]. A model was developed for a specially designed copper tool with optimized parameters to improve the MRR of a blind hole and results shows that there is 300% improvement in MRR [9]. TWR, EWR, and R_a are the output metrics that come from the machining input parameters. Due to the exceptional mechanical properties of metal matrix composites (MMCs), industrialists are currently looking for advanced uses for these materials. MMCs reduce the weight and cost of materials while improving products' strength and impact resistance [10, 11]. EDM is preferred over other unconventional machining methods such as chemical machining, LBM, and EBM in the industries due to its low cost and safety. EDM is widely used for intricate parts such as drilling cooling holes in turbine parts and cooling holes with improved quality [12]. Experiments were performed cryogenically treated brass and copper electrodes for Ti-6Al-4 V alloy workpiece to reduce the MRR, radial overcut and tool wear [13]. The effect of various dielectric fluids (kerosene, deionized water, EDM emulsion and water solution EDM fluid) were investigated for drilling cooling holes in a nickel alloy Inconel 738 [14]. Input parameters has been optimized for Inconel alloy using gray-based fuzzy logic technique for MRR and hole taper [15]. In EDM process dielectric fluid plays an important role and maintain electrical insulation between workpiece and tool. It also help in dissipating heat, preventing thermal damage during machining

[16, 17]. Kerosene (easily available and low cost but low flash point) and hydrocarbon oil (superior cooling and insulating properties) are used as a dielectric fluid [18–20]. Graphite is generally employed as electrode because of its high conductivity and low wear properties to machine alloys such as copper, Inconel, steel, titanium, and aluminum [21–24]. Brass is occasionally used for wire EDM, while copper gives higher conductivity and better surface finish; it may wear down more quickly than graphite. The need for materials with unique characteristics has significantly increased within recent years, especially in the advanced industrial sectors like aerospace, medical devices, and surgical instruments [25–29].

Various investigations have been performed to increase the machining efficiency EDD process. However, a deep understanding of the underlying issues and essential mechanisms remains limited, as they have not been systematically analyzed. This paper aims to amalgamate research findings on EDD processes and provide an overview of the current scenario of research in this field and findings are summarized in tabular form. Future research directions and latest developments in this field are discussed at the end of the study.

2 Fundamentals of EDD Process

In-house EDD is a significant technological breakthrough in machining, as it enables the processing of electrically conductive materials regardless of their hardness. The EDD process, in contrast to EDM, cuts over the stationary workpiece material using a rotating tool electrode [18]. Figure 1 displays the line diagram for the EDD process. In this process cathode (tool electrode) and the anode (workpiece) are physically separated (generally 0.01–0.5 mm) and gap is filled with dielectric fluid [19].

This fluid removes debris, cools the machining region, and acts as a conduit for electrical discharge. Both the tool and the workpiece receive high-frequency pulsed DC voltages between 20 and 120 V. A plasma channel is created and the dielectric fluid is ionized by the potential difference this voltage creates. As the electrons travel from the tool electrode to the workpiece, they carry stray electrons. The dielectric breaks down as a result of this flow, producing a plasma channel with extraordinarily high pressures and temperatures. Because of the intense heat and pressure in the plasma channel, localized melting occurs [20].

3 Study of Composite Materials

Composites are made by macroscopically combining distinct materials to achieve better qualities than single ones. Due to the improved qualities composites are used in aerospace, automotive, construction, marine, sporting goods, and infrastructure sectors. A balanced set of qualities for composite materials is required for a certain range of applications, which requires property optimization [22]. With the use of

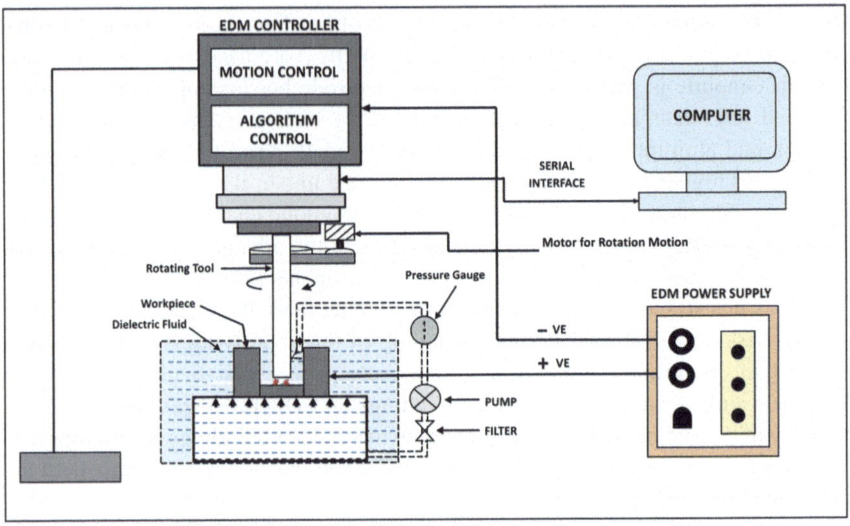

Fig. 1 Line diagram of EDD process

sophisticated machining techniques like EDM, EDD, die sinking, etc., the effects of various factors on MRR, R_a, and TWR have been investigated. In EDD, composite materials used are often chosen based on their conductivity and the specific application requirements [30]. MMCs are used due to their electrical conductivity and high strength. Aluminum Matrix Composites reinforced with particles like Silicon Carbide (SiC) or Alumina (Al2O3). Copper Matrix Composites are reinforced with materials like Graphite or Carbides. Carbon Fiber Reinforced Polymers is not conductive, it can be treated with a conductive layer or medium to enable EDD [31]. Ceramic Matrix Composites are used in high-temperature applications, SiC composites are common, but they need to be treated or coated with a conductive material to be used in EDD. Hybrid Composites can be a mix of metals and ceramics or metals and polymers, designed to offer the advantages of both materials for drilling with electrical discharge. Materials are selected based on their attributes or properties like current conductivity, strength, and heat resistance for specific uses in EDD processes [32].

4 Experimental Procedures

Many researchers have conducted experimental and theoretical studies on MMCs throughout the EDD process, looking at potential uses in Micro-electromechanical systems and advanced automated industries. Due to the intricacy of the EDD process mechanism for MRR, several experimental works have been conducted as opposed to theoretical studies [16–18]. The EDD setup using conventional EDM machines, which is a common strategy to minimize costs while expanding the functionality of existing equipment. Converting an EDM machine for use in drilling, grinding, or milling involves creating modular setups that can be easily mounted or removed from the machine's servo head [33]. For tool flexibility, electrode holders should accommodate electrodes of different shapes and sizes, enabling the machine to switch between applications without extensive reconfiguration [34]. In the EDD process, electrode wear is a major challenge, specifically during prolonged drilling operations. Continuous degrading of electrode affects machining accuracy and efficiency, requiring frequent electrode replacement. Developing materials and methods that reduce electrode wear without compromising performance is a critical aspect of in-house EDD development. Thermal expansion and the recast layer that develops on the workpiece make it challenging to maintain tight tolerances. Advanced control algorithms or cooling mechanisms might be needed to minimize these effects. In drilling, dielectric fluid helps carry away the eroded material (debris), maintain the spark gap, and cool the workpiece [35]. High-pressure flushing of dielectric fluid are essential to remove debris from the machining zone. Developing in-house high-pressure systems may be a challenge but is essential for improving hole quality. Discharge Energy Control systems should allow fine-tuning of discharge energy, current, and voltage to adapt the setup for many operations [36–38]. The common process variables namely voltage, current, pulse on-time, pulse off-time, dielectric flushing pressure, tool rotational speed and IEG and output characteristics like MRR, TWR, Overcut, and R_a are playing a significant role during EDD process for MMCs.

5 Critical Parametric Analysis

A thorough review of the literature revealed that numerous EDD technologies have been developed in response to the industry's expanding demands. As these technologies have advanced, machining time has decreased and machining efficiency has significantly increased. Because of this, EDD is more useful in industrial settings [39, 40]. Electrode composition and EDD configurations have a major impact on EDD performance. Selecting the appropriate electrode material and EDD settings is essential to reducing electrode wear and improving machining efficiency. By using electrodes of different shapes and aided EDD, the effectiveness of EDD is raised. Better machining performance can be attained by using different shaped electrodes (like step, side-cutter, grooved, and helical) allow for greater debris evacuation and prevent short circuits and arcing. Electrical (electrode tool gap, discharge voltage, peak current, polarity, pulse type, pulse length, and interval) and non-electrical (dielectric fluid's cleaning action, the electrode tool's rotation, and the characteristics of the workpiece) factors affect the EDD process [41–44]. Table 1 shows the parametric analysis of various materials based on research from the past 15 years.

Table 1 Summary of the critical study of EDD process

Input parameters with ranges	Output parameters	Workpiece	Dielectric	Findings	Ref.
Current: (1–9) A, pulse on-time: 10, 30, 50, and 90 (μs), duty factor: 10% to 90%, servo speed: (2–10) mm/s, water pressure: 20–60 kg/cm^3	MRR, ERR, Hole taper	Stainless steel (AISI 431)	Deionized water	As the servo rate increases, the MRR decreases. Additionally, as electrical current and servo rate increase, the hole's taper increases	[20]
Voltage gaps: 20, 30, 40, and 50 (V) dimensions of the electrode: 0.8, 1, 2, and 3 mm conductivity: 1142, 2116, 3074, 408 (μS/cm)	TWR and MRR	Inconel 718	Super dielectric fluid	The theoretical and experimental observation are used to create the machining technique with the super dielectric fluid	[45]
Current: 9, 12, 15, (A). Pulse on-time: 60, 90, and 120 (μs), pulse off-time: 15, 45, and 90 (μs) tool speed: 800–1200 rpm	MRR and R_a	Titanium alloy (Ti–6Al–4 V)	Deionized water	Investigation of the effects of EDD assisted by low-frequency vibration	[46]
Current applied: 2–5–8 (A); pulse on-time: 20–50–80 (μm), pulse off-time: 20–50–80 (μm) voltage: 40–60–80 V. Adjuster for gap voltage: 2–5–8	Taper ratio, hole dilatation, and specific drilling energy	Al 7075	Deionized water	0.88%, 1.69%, and 1.16% are the comparative deviations for SDE, DH, and TR, respectively	[47]
Time-sensitive pulse: 20–50–100 (μs), Pulse off-time: 20–40–60 (μs)	Effects of interaction on the geometrical	Inconel-718	Powder-mix-dielectric fluid	Circularity has improved as a result of the RSM-based model	[48]

(continued)

Table 1 (continued)

Input parameters with ranges	Output parameters	Workpiece	Dielectric	Findings	Ref.
Discharge: 25–30–35 V Flushing pressure: 0.5–0.75 (kg/cm^2)	Feature of holes				
Pulse on-time: 8–9–10 (μs); pulse off-time: 2–4–6 (μs) Current: 14–15–16 (A) Pressure of distilled water: 60–70–80 (kg/cm^2)	Circularity tapper and dilation of holes	Titanium alloy (Grade-5)	Deionized water	Results shows that the response variables for circularity were 28.57, hole taper was 14.28, and hole dilatation was 47.36	[49]
Pulse on-time: 3–6–9 (μs); pulse off-time: 3–5–7 (μs); Pressure during flushing: 3–5 (kg/cm^2)	MRR, TWR	AL 7075	Deionized water	Relationship between machining parameters and response variables examined in AL 7075 EDM drilling process	[50]
Current: 10 (A) Capacitance (mF): 1.2 Pulse on-time: 90 (ms); pulse off-time.90 ms; rotation of the electrode: 200 rpm	MRR, TWR, overcut, and taper angle	Al–Mg–Ti alloys	Deionized water	To improve the MRR and reduce the expansion of the hole and drilled energy, current study focused on improving EDD for the material SS304	[51]
Tool electrode: Copper 10 mm Dia Polarity: Straight Voltage: 80(V) Current: 10(A) Pulse on-time: 15 (μm) Interval time: 15 (μm)	MRR, TWR	Ti 6246	Kerosene	For varying drilling periods, improvements of up to 8.5% for MRR, 34.78% for TWR, and 30.16% for WR have been observed	[52]

6 Future Directions

Composite materials are difficult to machine due to their anisotropic and non-homogeneous structure as well as the high abrasiveness of their reinforcing components. This generally harms the workpiece and soon wears down the cutting tool. Standard machining operations like turning, drilling, and milling can be performed on composite materials if the right tool design and operating conditions are applied [53–55]. Because of their specific properties, composite materials can compete with conventional materials in many structural applications as long as they can be shaped into the appropriate shapes at a reasonable cost and performance level [56]. In the field of aerospace, automotive, automobile, and electronics are employed various composite materials for the increase of safety, compact electronic device, and sustainability.

The different composite materials significantly increase the efficiency of process parameters, surface finish, machining performances and reduce the surface defects during EDD process. The scopes of the various composite materials in the last few years are illustrated in Fig. 2. Research is also being done on the usage of different electrode materials, forms, sizes, and geometries. Although they are still in their infancy, tubular electrode instruments have demonstrated encouraging outcomes and need greater consideration. Performance parameters are the elements that gauge the effectiveness of the EDD process. Percentage of publication of the various output parameters (MRR, TWR, R_a, and Overcut) during the past few years are shown in Fig. 3.

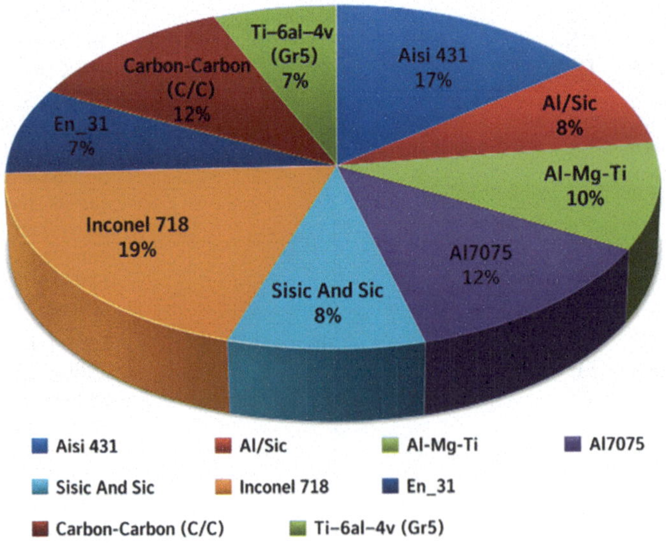

Fig. 2 Study on composite materials in the last few years

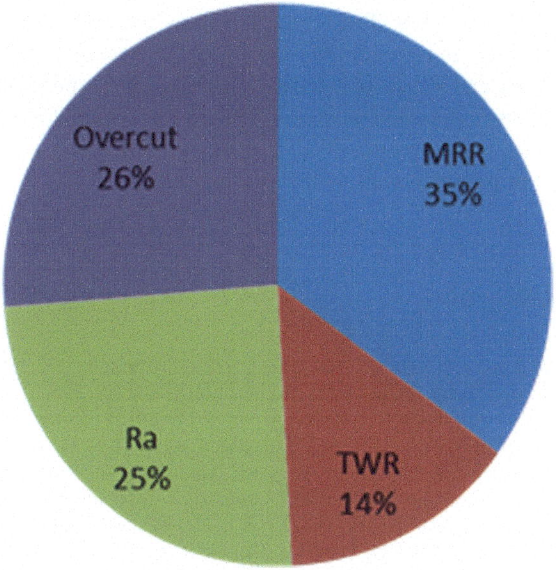

Fig. 3 Percentage of publication for different responses of EDD process

New cooling techniques, such as cryogenic cooling, have demonstrated promise in lowering TWR. Performance of EDD process can be improved by combining it with other procedures like magnetic force, rotating motion laser, or ultrasonic aid. With an emphasis on process optimization, material adaptability, tool innovation, and hybrid approaches, these points highlight the continuous developments in EDD technology.

7 Conclusion

This literature review focused on parametric Electrical Discharge Drilling (EDD) research on different composite materials. To examine the parametric research of EDD across different material compositions, more than 50 publications have been examined. Despite the promising outcomes of EDD, issues such as low MRR and high TWR still exist, especially with more modern industrial materials. When machining various stainless-steel grades, the discharge current is the primary factor affecting MRR. Fluid pressure, the speed at which the electrode tool rotates, and the time of pulse duration are further crucial factors. Pulse length and discharge current are the two main variables that impact the electrode wear rate (EWR). Similarly, Ra is improved, however, by shorter pulse on-time, longer pulse off-time, and lower pulsed current. Reduced tool wear ratios can produce deeper, more precise micro-holes, although further research is needed in this area. Based on a meta-analysis, the most work was done on composite materials that are electrically conductive. Moreover, the least amount of work was done on 10% Al–Mg–Ti, 8% AlSiC, and

7% Ti–6Al–4 V. The research reveals that 35% and 26% of effort has been done for MRR and overcut, but less effort has been made like 25% and 14% for studying the R_a and TWR, respectively. The review emphasizes the need for more study in process innovation, namely in the areas of parameter optimization to tackle persistent issues like Overcut, TWR, and R_a, as well as the exploration of hybrid approaches for improved results.

References

1. Mao X, Almeida S, Mo J, Ding S (2022) The state of the art of electrical discharge drilling: a review. Int J Adv Manuf Technol 121(5–6):2947–2969
2. Kumar S, Singh R, Batish A, Singh TP (2012) Electric discharge machining of titanium and its alloys: a review. Int J Mach Mach Mater 11(1):84–111
3. Kumar A, Singh VP, Singh RC, Chaudhary R, Kumar D, Mourad AHI (2024) A review of aluminum metal matrix composites: fabrication route, reinforcements, microstructural, mechanical, and corrosion properties. J Mater Sci 59(7):2644–2711
4. Darji Y, Patel D, Patel D, Ramesh R, Oza AD, Bhole KS, Shinde SM, Kumar M (2023) Experimentation with the EDM parameter through a full factorial technique and optimization using regression analysis with carbon nanotubes. Int J Interact Des Manuf. https://doi.org/10.1007/s12008-023-01263-2
5. Kumar P, Yadav SKS (2019) Multi-objective optimization of electrical discharge drilling (EDD) process using PCA-based grey relational analysis. Mater Today Proc 26(xx):2013–2020
6. Kachhap S, Singh A, Debnath K (2018) Electric discharge drilling of hybrid metal matrix composites using different tool electrodes. J Sci Ind Res 77(6):325–329
7. Yadav US, Yadava V (2017) Experimental investigation on electrical discharge diamond drilling of nickel-based superalloy aerospace material. Proc Inst Mech Eng Part B J Eng Manuf 231(7):1160–1168
8. Saini SK, Khaja SM (2023) Evolution of optimum parameters for EDD in tool steel. E3S Web Conf 430:1–5
9. Nastasi R, Koshy P (2014) Analysis and performance of slotted tools in electrical discharge drilling. CIRP Ann-Manuf Technol 63(1):205–208
10. Mao X, Wu G, Tran M, Yi S, Ding S (2024) Electrical discharge drilling of blind holes with injection flushing dielectric and stepped electrodes. Int J Adv Manuf Technol 132(1–2):495–511
11. Harane PP, Wojciechowski S, Unune DR (2022) Investigating the effect of different tool electrodes in electric discharge drilling of Waspaloy on process responses. J Mater Res Technol 20(xx):2542–2557
12. Janmanee P, Muttamara A (2011) A study of hole drilling on stainless steel AISI 431 by EDM using brass tube electrode. Int Trans J Eng Manag Appl Sci Technol 2(4):471–481
13. Shastri RK, Mohanty CP, Jadhav PS (2019) Study of the effect of cryogenically treated tools during rapid drill electro-discharge machining of Ti-6Al-4V alloy. Mater Today Proc 18:3122–3127
14. Li C, Xu X, Li Y, Tong H, Ding S, Kong Q, Zhao L, Ding J (2019) Effects of dielectric fluids on surface integrity for the recast layer in high-speed EDM drilling of nickel alloy. J Alloys Compd 783:95–102
15. Arif T, Sharma A (2024) Experimental investigation in sustainable electric discharge drilling of Inconel alloy using grey-based fuzzy logic approach. Proc Inst Mech Eng Part E J Process Mech Eng. https://doi.org/10.1177/09544089231221541
16. Rampal G, Kumar G, Rangappa SM, Siengchin S, Zafar S (2022) A review of recent advancements in drilling of fiber-reinforced polymer composites. Compos Part C Open Access 9(xx):100312

17. Yudin AS, Zhurkov MY, Martemyanov SM, Datskevich SY, Vazhov VF (2019) Electrical discharge drilling of granite with positive and negative polarity of voltage pulses. Int J Rock Mech Min Sci 123(xx):104058
18. Pelanconi M, Bottacin S, Bianchi G, Koch D, Colombo P, Ortona A (2024) High-strength Si–SiC lattices prepared by powder bed fusion, infiltration-pyrolysis, and reactive silicon infiltration. J Am Ceram Soc 107(7):4436–4450
19. Dhaker KL, Bhilala K, Sharma HK, Bhil P (2023) Experimental investigation of hole geometry during electric discharge drilling of aerospace material sheet. Mater Today Proc 78:570–579
20. Govindan P, Joshi SS (2010) Experimental characterization of material removal in dry electrical discharge drilling. Int J Mach Tools Manuf 50(5):431–443
21. Weiwen X, Junqi W, Wansheng Z (2018) Break-out detection for high-speed small hole drilling EDM based on machine learning. Procedia CIRP 68(xx):569–574
22. Singh A, Kachhap S, Kumar R (2019) 14-electro-discharge drilling of metal matrix composites, In: Woodhead publishing series in composites science and engineering, hole-making and drilling technology for composites. Woodhead Publishing, pp 193–210
23. Singh K, Singh H, Vardhan S, Mohan S (2020) Mechanical study of Al 7050 and Al 7075 based metal matrix composites: a review. Mater Today Proc 43(xx):673–677
24. Jeevamalar J, Kumar SB, Ramu P, Suresh G, Senthilnathan K (2021) Investigating the effects of copper cadmium electrode on Inconel 718 during EDM drilling. Mater Today Proc 45(xx):1451–1455
25. Abu Qudeiri JE, Saleh A, Ziout A, Mourad AHI, Abidi MH, Elkaseer A (2019) Advanced electric discharge machining of stainless steels: assessment of the state of the art, gaps and future prospect. Materials 12(6):xx
26. Luo R (2002) Fabrication of carbon/carbon composites by an electrified preform heating CVI method. Carbon 40(11):1957–1963
27. Kyratsis P, Tzotzis A, Manavis A (2021) Computational design and digital fabrication. In: Kumar S, Rajurkar KP (eds) Advances in manufacturing systems, Lecture notes in mechanical engineering. Springer, Singapore, pp 1–16
28. Baldan R, Rocha R, Tomasiello R, Nunes C, Costa A, Barboza M, Coelho G, Rosenthal R (2013) Solutioning and aging of MAR-M247 nickel-based Superalloy. J Mater Eng Perform 22:2574–2579
29. Park K, Park J, Kwon H (2017) Fabrication and characterization of Al-SUS316L compo-site materials manufactured by the spark plasma sintering process. Mater Sci Eng A 691:8–15
30. Hu B, Mo JPT, Ding S, Brandt M (2016) Model-driven design of a fast material removal electrical discharge machine. Cogent Eng 3(1):1–17
31. Uhlmann E, Polte J, Streckenbach J, Dinh NC, Yabroudi S, Polte M, Börn-stein J (2022) High-performance electro-discharge drilling with a novel type of oxidized tool electrode. J Manuf Mater Process 6(5):113
32. Diver C, Atkinson J, Helml HJ, Li L (2004) Micro-EDM drilling of tapered holes for industrial applications. J Mater Process Technol 149(1–3):296–303
33. Singh S, Maheshwari S, Pandey PC (2004) Some investigations into the electric dis-charge machining of hardened tool steel using different electrode materials. J Mater Process Technol 149(1–3):272–277
34. Wang CC, Yan BH (2000) Blind-hole drilling of Al2O3/6061Al composite using rotary electro-discharge machining. J Mater Process Technol 102(1):90–102
35. Schumacher BM (1990) About the role of debris in the gap during electrical discharge machining. CIRP Ann-Manuf Technol 39(1):197–199
36. Ayesta I, Flaño O, Izquierdo B, Sanchez JA, Plaza S (2016) Experimental study on debris evacuation during slot EDMing. Procedia CIRP 42(xx):6–11
37. Wang J, Han F (2014) Simulation model of debris and bubble movement in electrode jump of electrical discharge machining. Int J Adv Manuf Technol 74(5–8):591–598
38. Mohd Abbas N, Solomon DG, Bahari MF (2007) A review on current research trends in electrical discharge machining (EDM). Int J Mach Tools Manuf 47(7–8):1214–1228

39. Manikandan P et al (2021) Influence of mixed dielectric fluids on material removal performance during electric discharge machining process. Mater Today Proc 51(xx):720–722
40. Gostimirovic M, Kovac P, Skoric B, Sekulic M (2011) Effect of electrical pulse parameters on the machining performance in EDM. Indian J Eng Mater Sci 18(6):411–415
41. Cao H-T, Ho J-R, Tung P-C, Tsui H-P, Lin C-K (2025) Characterization of ma-chined surface in semi-conductive SiC wafer subjected to micro-EDM drilling. Mater Sci Semicond Process 187:109118
42. Sun Y, Sun Y, Junyi D, Yin J, Li S, Gong S, Sun M, Liu M, Li X (2024) Review on role of electrical discharge drilling methods in fabricating micro holes: formation mechanism, defects characterization and mitigation strategies. Arch Civ Mech Eng 24:143
43. Yıldız Y (2018) Experimental investigation of machining accuracy in electrical discharge drilling. Pamukkale Univ J Eng Sci 24(1):69–75
44. Muthuramalingam T, Mohan B (2015) A review on influence of electrical process parameters in EDM process. Arch Civ Mech Eng 15(1):87–94
45. Giridharan A, Oommen J, Srisailesh VS, Mohan R (2019) Electrical discharge drilling of micro-hole on Inconel 718 using rotary tubular electrode. Mater Today Proc 22:1723–1730
46. Tanjilul M, Wee Keong DN, Senthil Kumar A (2021) Super dielectric based EDM process for drilling of Inconel 718. Mater Manuf Process 36(3):341–350
47. Zhang M, Cai T, Lu W, Zheng S, Zhang Y, Zhang Y (2023) Fabrication of Ti-6Al-4V alloy powder by a novel sintering-deoxygenation process. J Alloys Compd 935:168113
48. Wang X, Qu N, Fang X, Li H (2016) Electrochemical drilling with constant electrolyte flow. J Mater Process Technol 238:1–7
49. Jain A, Pandey AK (2019) Modeling and optimizing of different quality characteristics in electrical discharge drilling of titanium alloy (Grade-5) sheet. Mater Today Proc 18:182–191
50. Khanna R, Kumar A, Garg MP, Singh A, Sharma N (2015) Multiple performance characteristics optimization for Al 7075 on electric discharge drilling by Taguchi grey relational theory. J Ind Eng Int 11(4):459–472
51. Eyercioglu O, Gov K (2021) The effect of magnesium content on drilling of Al-Mg-Ti alloy by hole electrical discharge machining process. Proc Inst Mech Eng Part B J Eng Manuf 235(1–2):125–133
52. Gill SS, Singh J (2010) Effect of deep cryogenic treatment on machinability of titanium alloy (Ti-6246) in electric discharge drilling. Mater Manuf Process 25(6):378–385
53. Ablyaz TR, Shlykov ES, Muratov KR (2021) Improving the efficiency of electrical discharge machining of special-purpose products with composite electrode tools. Materials 14(20)
54. Bhuyan BK, Bhuyan P, Mishra S (2020) Modeling and response optimization of traveling wire electro-chemical spark machining of borosilicate glass using hybrid approach. J Adv Manuf Syst 19(3):425–447
55. Matharou GS, Bhuyan BK (2021) Parametric analysis of electric discharge machining of hybrid composite materials. In: Sharma BP, Rao GS, Gupta S, Gupta P, Prasad A (eds) Advances in engineering materials, Lecture notes in mechanical engineering. Springer, Singapore, pp 329–339
56. Kachhap S, Singh A, Kumar S (2019) Performance evaluation of different electrode geometries in electric discharge drilling of MMCs. Int J Mech Eng Robot Res 8(4):531–535

An Analysis of E-Procurement Implementation Obstacles in Indian Industries: A DEMATEL Approach

Lipika Mishra, Sushanta Tripathy, and Deepak Singhal

Abstract Procurement is a core function in any business. Currently with changing dynamics, e-procurement is shaping up as a new emerging supply chain dimension to gain a competitive edge over the industries. The different e-procurement implementation obstacles have been found after a critical review of the past literatures. The concerned research is addressed to decode the e-procurement obstacles in the Indian context. A robust DEMATEL MCDM technique is employed to explore the obstacles, their importance, and their inherent nature. The results show that policy and regulatory obstacles are the most prominent ones followed by supply chain obstacles and data management obstacles as the top three among the seven decoded obstacles. Further, technical, social cultural, organizational, and sustainability obstacles are "effect" natured which reflects they cannot be resolved directly by attacking them. This research would aid in propelling the e-procurement implementation and act as a sturdy base for other nations new to implementing them.

Keywords Procurement · DEMATEL MCDM · E-procurement

1 Introduction

Procurement is a core function in any manufacturing and trade business. Any industry that imbibes procurement, a core competency, helps it to strive and thrive in the market and develop a competitive edge over its competitors [1]. In 2004, Hartmann vastly explored the effect of IT on procurement concerns and observed that it can offer smoother and faster process flow, efficient information distribution, increased transparency, better control, decentralization of tasks and decisions, and would support internal processes as well as involve coordination between business partners [2].

Although e-procurement has gained prominence over the past decade, the nation-specific approaches towards e-procurement and its obstacles are dispersed in nature.

L. Mishra · S. Tripathy (✉) · D. Singhal
Mechanical Engineering, KIIT Deemed to be University, Bhubaneswar, India
e-mail: sushant.tripathy@gmail.com

A. Biswas et al. (eds.), *Recent Advancements in Artificial Intelligence and Optimization*, Lecture Notes in Mechanical Engineering,
https://doi.org/10.1007/978-981-95-9268-5_14

The mitigation is formidable in their own contexts, and an overall approach to e-procurement management challenges in developing countries is yet to be explored. This research presents a fresh view of e-procurement implementation obstacles and their prioritization along with an analysis to decode the path of mitigation approach in the Indian context. To implement mitigation strategies, the obstacles are to be identified and prioritized on one hand and segregated to channel pathways to resolve them. The lack of an integrated analysis of e-procurement obstacles in the industrial context of India intrigued the authors to begin the research to identify the obstacles in e-procurement implementation. Further, the obstacles are prioritized through ranking and hence a clear picture of understanding evolves for the stakeholders. This led the authors to propose the research questions (RQs) for this research.

RQ1: To identify the obstacles in e-procurement implementation in the Indian context (?)

RQ2: To prioritize/rank the challenges identified through weightage to decode the importance of each of the obstacles (?)

RQ3: To decode cause and effect categories of obstacles to trace the focus on implementation of e-waste initiatives (?)

RQ1 was addressed through the vast literature survey, experts' opinions, and the formidable MCDM methods, viz. DEMATEL was used to rank the challenges through weightage and build the model (**RQ2** and **RQ3**).

The framework of the paper is as follows: Sect. 2 showcases the literature pertaining to e-waste barriers resulting in Table 1; Sect. 3 delves into the research design/methodology section which showcases data collection methods and describes the MCDM technique, viz. DEMATEL; Sect. 4 involves the results; Sect. 5 comprises the discussion of findings; Sect. 6 showcases the practical implications; and Sect. 7 concludes the research.

Table. 1 Implementation obstacles of e-procurement identified in the literature

Codes	Obstacles in e-procurement
TO	Technical obstacles
OSO	Organizational and sustainability obstacles
PRO	Policy and regulatory obstacles
SCO	Supply chain obstacles
FO	Financial obstacles
SACO	Social and cultural obstacles
DMO	Data management obstacles

2 Literature Review

The motivation for authors ignited to build on this research involving e-procurement obstacles as the supply chain and its strategies were revamped owing to the post-pandemic (COVID-19) scenario. It initiated through a systematic literature analysis to get a picture of digitization in e-procurement obstacles pertaining to the Indian context.

The modern reincarnation of digitization of e-procurement as depicted by Nicoletti and Nicoletti [3], when they addressed the future of procurement as procurement 4.0. The digital procurement like bitcoin is supported by I 4.0 technology, viz., blockchain, for digital security, which also has been utilized for procurement finance and smart contracts. Issah and Ackah [4] concluded that electronic procurement has a positive and significant effect on procurement performance. The authors reviewed several recent articles Bienhaus and Haddud [5]; Ibrahim et al. [6]; Winnata and Shamangun to finalize the obstacles in the implementation of e-procurement in the Indian context. These obstacles are listed along with their codes in Table 1.

The seven obstacles are elaborated in paragraphs below:

(i) TO: Technical Obstacles

With digitization of procurement as the new standard, technology is the most fundamental issue which would form the basis for any other improvement required towards system building. It is a prerequisite to overcome before any profound organizational and advanced hindrances can be dealt with. An integrative balance is the key to be addressed towards alleviating technical obstacles for initiating e-procurement [7].

(ii) SCO: Supply Chain Obstacles

Supply chain management has to be revamped to adjust to e-procurement. This includes reduction in number of suppliers, consolidating contracts and suppliers, and to centralize supply chain operations. Overall, a very dynamic challenge or obstacle is to be alleviated before we realize e-procurement in practice in any firm.

(iii) PRO: Policy and Regulatory Obstacles

A policy should be stringent as well as flexible enough to cater the variations in executing in various types of sectors/industries for maximizing results. Thus, a coherent regulatory framework is the need-of-the-hour to mitigate the constraints of e-procurement commencement. Presently, it is an intense obstacle with very few expertize and practical know-how or experience to chalk-out an all-inclusive policy for e-procurement [8].

(iv) DMO: Data Management Obstacles

With data as the new fuel, driving the industrial decision-making processes, data management, in general, is cumbersome. To the specificity of e-procurement implementation challenges, the play between technology and data management, in the dynamically driven supply chain to e-procurement environment, a balance between the two becomes absolutely essential to overcome this obstacle and harness its power to escalate performance of any firm.

(v) FO: Financial Obstacles

For any industry, entering into a new venture, particularly e-procurement has to invest financially trying to balance the complexities in supply chain and encompassing the scale of operations involved. This hindrance is crucial to be addressed to initiate even the other obstacles identified in this research [9].

(vi) SACO: Social and Cultural Obstacles

The social and cultural aspects of the industry imbibing e-procurement should be proactive to infuse into their work environment, the impression of the issue to the workforce so as to transmit awareness and an orientation through thinking, brainstorming, and discussions along with awareness campaigns. This would be pre-cursor to flexible acclimatization towards shaping-up of industry's workforce [10].

(vii) OSO: Organizational and Sustainability Obstacles

This obstacle is the culmination of all the other six barriers identified in this research. The alleviation of OSO can only be achieved when the other six of them have been resolved successfully. It is the goal to be achieved for initiation of e-procurement implementation process [11].

3 Methodology

A real-life system has several components with numerous intertwining relationships that are difficult to visualize due to its inherent complexities. Gabus and Fontela [12] formulated a powerful MCDM technique that eventually crystallized the interrelationships between components of a system and analyzed the "structural fiber" of an industrial system having multiple components [13].

This research is utilized to establish the cause-and-effect obstacles in e-procurement issues after the obstacles are identified and validated through experts' opinions. The following Sect. 3.1 and 3.2 address the methods of data collection and the methodology to decode the data, i.e., experts' opinions towards a clear picture of obstacles in e-procurement implementation.

3.1 Data Collection

Systematic literature scrutinized by the authors led to the detection of the obstacles which was validated through the experts' opinions through a questionnaire using the 5-point Likert scale which reads as: 0: NI; 1: LI; 2: I; 3: HI; 4: VHI.; where I: influence; H: High; L: Low; V: Very; N: No. The number of experts being four, seven obstacles were decoded.

3.2 DEMATEL MCDM Technique

The DEMATEL MCDM can be executed through the following steps:

D-1: Collecting experts' opinions to generate the direct relation matrix by taking the average of each cell for "n" number of experts.

$$\text{DR} = \{a_{ij}\} \tag{1}$$

where DR = n x n positive averaged scale values matrix;
a_{ij} = explicit intensity of barrier i on barrier j;
when $i = j \Rightarrow a_{ij} = 0$ (the diagonal elements).

D-2: Normalized version of the direct matrix is obtained

$$\text{Nor} = \{n_{ij}\} \tag{2}$$

where each of n_{ij} = Each cell of DR/ max $\left\{\sum_{i=1}^{n} a_{ij}, \sum_{j=1}^{n} a_{ij}\right\}$

D-3: Obtain the total relation Matrix [TR]

$$[\text{TR}] = \text{Nor}\ x\ (I - \text{Nor}) - 1 \tag{3}$$

where I = n x n identity matrix.

Each element of $(\text{TR})_{ij}$ express the implicit effects of obstacle i on j; thus, exemplifying the pairwise relationship of all obstacles in totality per cell of [TR].

D-4: Summation of rows (sr) and columns (sc) of [TR]

$$sr = \sum_{1 \le j \le n} K(i, j) \tag{4}$$

$$sc = \sum_{1 \le i \le n} L(i, j) \tag{5}$$

The (sr + sc) and (sr−sc) values showcase the "importance/prominence" of each obstacle and "causal or effect" among them, respectively.

If (sr−sc) = positive ⇒ Causal obstacles
= negative ⇒ effect obstacles

D-5: Scatter Plot {with (sr+sc) as x-axis and (sr−sc) as y-axis}. It leads to the visualization of causal and effect obstacles and order of importance.

4 Results

The DEMATEL method as described in Sect. 3.2 has been utilized to a number of expert opinions (n = 4) who's each cell average value over 7×7 matrix forms the average matrix Table 2. This is the prerequisite to DEMATEL calculations.

The maximum sum of each row and column is chosen to be divided through each of the (7 × 7) 49 cells. The maximum is row sum of 18.5.

Max ($\sum$ROW, $\sum$COLUMN) = 18.5; so, each cell (7 × 7) is to be multiplied by (1/18.5) = 0.054.

The normalization matrix (N) is represented in Table 3.

The total relation matrix (Table 4) was calculated through a formula (Sect. 3.2) evaluated in Microsoft Excel 2010. Further, row sum and column sum are calculated and two other columns are generated by adding (sr+sc) and subtracting (sr−sc) them, respectively.

A threshold value "ά" is generated by summing all the values of obstacles (7 × 7 = 49) in cells of the total relation matrix and dividing the obtained value by the number of cells, i.e., 49. The value of "ά" is 0.84.

Table. 2 Average matrix

Obstacles	TO	OSO	PRO	SCO	FO	SACO	DMO	Row sum
TO	0	2.5	1.25	1.5	2.5	2	2.75	12.5
OSO	2.25	0	2.5	2	2	3	1	12.75
PRO	2.25	3	0	4	2.5	3.25	3	18
SCO	1	3	4	0	4	3.5	3	18.5
FO	4	3.5		2	0	3	3	17.5
SACO	1	3	4	2.75	1	0	1.5	13.25
DMO	4	3	4	4	2.25	1.25	0	18.5
Col.sum	14.5	18	17.75	16.25	14.25	16	14.25	

Table. 3 Normalization matrix

1/ max sum	0.054	0.054	0.054	0.054	0.054	0.054	0.054
Obstacles	TO	OSO	PRO	SCO	FO	SACO	DMO
TO	0	0.135	0.0675	0.081	0.135	0.108	0.1485
OSO	0.1215	0	0.135	0.108	0.108	0.162	0.054
PRO	0.1215	0.162	0	0.216	0.135	0.1755	0.162
SCO	0.054	0.162	0.216	0	0.216	0.189	0.162
FO	0.216	0.189	0.108	0.108	0	0.162	0.162
SACO	0.054	0.162	0.216	0.1485	0.054	0	0.081
DMO	0.216	0.162	0.216	0.216	0.1215	0.0675	0

Table. 4 Total relation matrix

Obstacles	TO	OSO	PRO	SCO	FO	SACO	DMO	r.sum	col.sum	r + c	r-c
TO	0.54	0.78	0.72	0.68	0.65	0.7	0.66	4.74	5.34	10.08	−0.6
OSO	0.64	0.67	0.78	0.71	0.64	0.75	0.59	4.78	6.59	11.37	−1.81
PRO	0.86	1.07	0.93	1.04	0.88	1	0.89	6.66	6.55	13.21	0.11
SCO	0.84	1.1	1.14	0.89	0.96	1.04	0.92	6.89	6.06	12.95	0.83
FO	0.89	1.03	0.96	0.9	0.71	0.94	0.84	6.27	5.37	11.63	0.9
SACO	0.62	0.85	0.89	0.79	0.64	0.66	0.65	5.1	6.03	11.13	−0.93
DMO	0.95	1.09	1.12	1.06	0.89	0.94	0.77	6.83	5.32	12.15	1.51

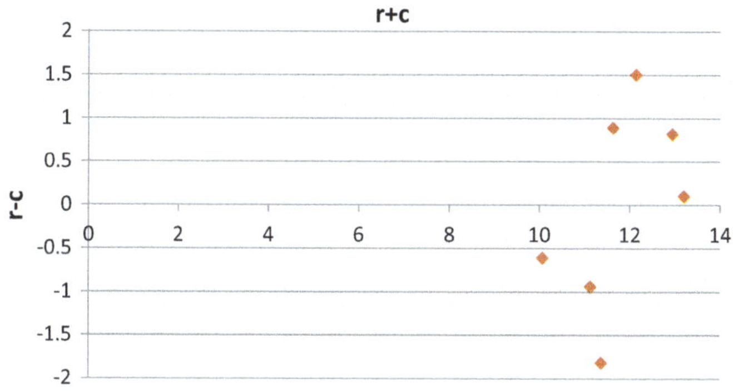

Fig. 1 Prominence-causal diagram

A scatter plot is generated with (sr+sc)=r+c as x-axis and (sr−sc) = r−c as y-axis. This is depicted in Fig. 1 below. The (sr + sc) value magnitude for each obstacle reflects the importance of obstacles, while (sr−sc) values instills categorization of the nature of obstacles as cause-and-effect types.

5 Discussion

The authors feel to have diligently worked upon the objectives of this research and found the results shown in Sect. 4 of this manuscript. The following paragraphs delve deep into extracting the insights as discussed below:

The average matrix exhibits the average aggregate of all the 4 experts' opinions obtained through a closed-ended questionnaire. It is a prerequisite to perform DEMATEL to generate the normalization matrix (N), (I-N)-1 matrix, and the total relation matrix (N* (I-N)-1). The total relation matrix leads to (sr + sc) and (sr−sc) columns which generate the prominence-causal diagram in the form of a scatter plot.

The total relation matrix's (sr + sc) value determines the degree of importance of each obstacle towards e-procurement implementation, while the (sr−sc) determines the nature of each obstacle as causal or an effect.

The importance of obstacles aligned in descending order is showcased below [(r + c) values]:

$$\begin{aligned} &\text{PRO } (13.21) > \text{SCO } (12.95) > \text{DMO } (12.15) \\ &> \text{FO } (11.63) > \text{OSO } (11.37) > \text{SACO } (11.13). \\ &> \text{TO } (10.08) \end{aligned}$$

It shows that policy and regulatory obstacles have the highest importance followed by Supply chain obstacles, data management, financial, organizational, and sustainability obstacles with social cultural, and technological hurdles being the least important.

The (r−c) values categorized the obstacles as cause–effect factors depicted as.

PRO, SCO, FO, and DMO: Causal obstacles (+ ve values)

TO, SACO, and OSO: Effect factors (− ve values)

It demarcates that the least important factors (last 3) are the effect factors. This conforms to the objective of the research that the causal obstacles are prominent to be addressed and effect factors cannot be addressed directly, so least important in this context. Further, in a general sense, organizational and sustainability have the highest negative value indicating that it is the ultimate goal/effect to be attained.

6 Practical Implications

The research results would be an eye-opener to all stakeholders, viz., practitioners, managers, industry specialists as well as policymakers to explore the structure and nature of obstacles and thus would help them to expedite while enduring for an endeavor to e-procurement implementation. Further, the top management can be convinced to prioritize on the issue based on a concrete picture which this research provides about the structure and nature of obstacles. This initiation or "where to start" issue would be resolved and an action plan can be chalked out to propel towards achieving this business transformation goal.

7 Conclusions

The authors in this research have attempted to explore into the obstacles of e-procurement in Indian context. To accomplish it, three research questions (RQs) were posed and a robust MCIDM technique DEMATEL was employed to obtain a clear view on the nature of the obstacles and their importance.

The results showed that policy and regulatory obstacles, supply chain obstacles and data management obstacles are the most important hindrances to e-procurement implementation in Indian context. Further, technical obstacles, Social and Cultural obstacles, and organizational and sustainability obstacles are the effect factors which implies that they cannot be alleviated directly by attacking them rather by resolving the cause obstacles would help to tackle them. This research can be expanded to different regions with scope for new regional niche-specific barriers or opinions from experts from variety of industries may give a fresh vision to overall divergence in results. This research will act as a sturdy base for future research as well as a guiding force to all nations trying to decode the e-procurement obstacles around the globe.

References

1. Tanner C, Wolfle R, Schubert P, Quade M (2008) Current trends and challenges in electronic procurement: an empirical study. Electron Mark 18(1):6–18
2. Dooley K, Purchase S (2006) Factors influencing e-procurement usage. J Public Procure 6(1/2):28–45
3. Nicoletti B, Nicoletti B (2018) The future: procurement 4.0. In: Agile procurement: volume II: designing and implementing a digital transformation, pp 189–230
4. Issah O, Ackah D (2024) The role of organizational culture as a moderating factor in the relationship between e-procurement adoption, procurement performance, and value for money. Afr J Procure, Logist & Supply Chain Manag 7(9):1–23
5. Bienhaus F, Haddud A (2018) Procurement 4.0: factors influencing the digitisation of procurement and supply chains. Bus Process Manag J 24(4):965–984
6. Ibrahim M, Karollah B, Mahdani R (2023) The effect of supply chain innovation and e—procurement implementation on supply chain performance of manufacturing organization. Uncertain Supply Chain Manag 11(2):698–706
7. Celestin M, Sujatha S (2024) Digital procurement transformation in EAC: opportunities and challenges. Indo Am J Multidiscip Res Rev 8(2):78–92
8. Bajpai M, Prasad N, Tripathi A (2024) Modernization of business: roles of finance in business modernization. J Inform Educ Res 4(2):1829–1842
9. Jama EM, Mwanza BG, Mwanaumo EM (2024) E-procurement adoption barriers encountered by small and medium-sized enterprises (SMEs) in the Republic of South Sudan. Afr J Commer Stud 4(1):48–68
10. Mose JM, Njihia JM, Magutu PO (2013) The critical success factors and challenges in e-procurement adoption among large scale manufacturing firms in Nairobi, Kenya
11. Rejeb A, Rejeb K, Appolloni A, Seuring S (2024) Public procurement research: a bibliometric analysis. Int J Public Sect Manag 37(2):183–214
12. Gabus A, Fontela EJBGRC (1972) World problems, an invitation to further thought within the framework of DEMATEL, vol 1, 8. Battelle Geneva Research Center, Geneva, Switzerland, pp 12–14
13. Mohapatra B, Tripathy S, Singhal D (2023) A sustainable solution for lean barriers through a fuzzy DEMATEL methodology with a case study from the Indian manufacturing industry. Int J Lean Six Sigma 14(4):815–843
14. Cao F, Li R, Guo S (2024) Rhetoric and reality of public-private partnerships in China: a sustainable public procurement perspective. Socio-Econ Plan Sci 92:101852
15. Hartmann E, Ritter T, Gemuenden HG (2004) Determining the purchase situation: a comprehensive view on customer behaviour. J Cust Behav 3(3):335–349

16. Meschini S, Pattini G, Andreis E, Scomparin L, Di Giuda GM (2024) Identification of indicators and digital tools to enhance civic monitoring in public procurement. J Leg Aff Disput Resolut Eng Constr 16(1):05023008
17. Soong KK (2025) Malaysian SME employees accelerating into digital procurement usage during the COVID-19 pandemic. Int J Procure Manag 22(1):1–21

Friction Welding Thermal Modeling of 15 CdV6 Alloy Steel

M. S. Krishnalal, E. S. Mahendra Raj, P. Anchana, P. M. Ajith, and Sunilkumar Dhasan

Abstract Friction welding has gained significant attention in recent years due to its efficiency and reliability, particularly in the context of solid-state metal joining techniques. This project focuses on the thermal modeling of rotary friction welding for 15CdV6 steel, a low-carbon alloy known for its high yield strength, toughness, and excellent weld ability. The primary objective is to simulate the temperature distribution within the weld zone under varying input parameters using ANSYS APDL software. Simulations were conducted for four different frictions and upsetting pressures: 30, 50, 70, and 90 MPa. The results provided insights into the maximum temperature attained during the welding process. Experimental validation was carried out and the results of peak temperature increase with higher friction pressures and it is consistent with the simulation results. The error between the simulated and experimental results was within 5%, highlighting the accuracy of the thermal model developed. This study contributes to the limited literature on friction welding of 15CdV6 steel and offers a reliable approach for predicting weld temperatures using numerical simulations.

Keywords Friction welding · 15CdV6 alloy steel · Thermal modeling · ANSYS APDL · Validation

1 Introduction

Welding is a fundamental metal joining process extensively utilized in manufacturing industries for its efficiency and reliability [1]. It can be broadly categorized into two primary types: solid-state welding [2] and fusion welding [3] Fusion welding involves

M. S. Krishnalal · S. Dhasan (✉)
Department of Mechanical Engineering, National Institute of Technology Silchar, Silchar, Assam, India
e-mail: sunilkumar@mech.nits.ac.in

E. S. Mahendra Raj · P. Anchana · P. M. Ajith
College of Engineering, Trivandrum, Kerala, India

A. Biswas et al. (eds.), *Recent Advancements in Artificial Intelligence and Optimization*, Lecture Notes in Mechanical Engineering,
https://doi.org/10.1007/978-981-95-9268-5_15

the melting and subsequent solidification of the parent metal to create a bond, while solid-state welding relies on heating the metal to a plastic state, followed by the application of external pressure to achieve bonding. Fusion welding is known for its narrow heat-affected zone and minimal defects, making it suitable for various precision applications. Among solid-state welding techniques, rotary friction welding stands out as a non-conventional process that generates heat through mechanical friction between two surfaces in relative motion [4]. This heat, combined with a lateral force called upsetting pressure, plastically deforms the material and facilitates coalescence. Notably, no melting occurs during the process, making it more of a forging technique than a traditional welding method. Rotary friction welding is particularly effective for joining cylindrical work pieces [5]. The process typically involves rotating one work piece using a lathe machine while the other is held stationary and advanced toward the rotating part. The frictional heat generated at the interface brings the material to a plastic state, after which the rotation ceases and the application of upsetting pressure results in a strong metallic bond [6]. The focus of this project is the thermal modeling of friction welding for 15CdV6 steel, a low-carbon, high-strength, low-alloy steel that exhibits excellent yield strength, toughness, and weldability, making it an ideal material for demanding applications [7]. It is classified as low-carbon bainitic steel and shows a high strength-to-weight ratio. The alloy's composition, which includes significant concentrations of molybdenum and vanadium (each less than 1%), along with a fourfold increase in chromium content, contributes to its superior mechanical properties. With alloying elements making up less than 5% [8] of the steel's total composition, 15CdV6 is categorized as a low-alloy steel. Its applications include rocket motor casings [8], High-altitude Escape Solid Motor [9], solid rocket boosters, jet engines, nuclear reactors [10], pressure vessels [11], motorsport components [12], and more. This material is especially valuable in high-stress environments where strength and toughness are essential. Notably, welding this material can be achieved without requiring subsequent heat treatment and with minimal loss of mechanical properties. However, there is limited published literature on the friction welding process for 15CdV6 steel. Therefore, this study aims to model the thermal behavior of the weld zone using ANSYS APDL software by simulating temperature distribution for varying input parameters [13]. The objective is to determine the maximum temperature attained during welding and to validate the simulated results through experimental measurements using four sets of readings.

2 Materials and Methods

The experimental setup involved preparing cylindrical rods of 15CdV6, each 16 mm in diameter and 65 mm in length, by cutting and cleaning the specimens with acetone to ensure the weld surfaces were free of contaminants. For the friction welding process, one rod was fixed on a chuck and rotated at 1500 rpm, while the other rod was held stationary. Mechanical contact was made between the two parts, and in the first phase of welding, friction pressure was applied for a set duration to generate heat at the contact surface, bringing it to a plastic state. Once the desired temperature was reached, rotation was stopped, and additional upsetting pressure was applied to ensure proper joining of the two parts. The chemical and the physical properties were given in Tables 1 and 2.

The thermal modeling of this process was conducted in ANSYS APDL. Boundary conditions are crucial factors for simulation or modeling. Several key conditions must be applied to ensure accurate results. First, the parallel surface of the friction surface on the non-rotating side must be assigned as totally constrained. Second, friction pressure is applied to the parallel surface of the friction surface on the rotating side. Once the rotation stops, the upsetting pressure is applied to the same parallel surface. Additionally, during the rotation phase, a heat flux equation is applied to the weld zone. Convective and radiation losses are accounted by providing value to the convective heat transfer coefficient h and surface emissivity ε. For the simulation, $h = 25\ W/m^2$ and $\varepsilon = 0.6$. These conditions collectively play a vital role in the effectiveness of the simulation. Fine meshing was applied near the friction surface to ensure accurate temperature distribution, with the goal of simulating the peak temperatures under varying friction and upsetting pressures. Geometry and the meshing is in Fig. 1.

For experimental validation, trials were conducted at Annamalai University in Chidambaram Fig. 2, where the peak temperature at the weld interface was monitored using an infrared thermometer [15]. These experimental results were then compared with the peak temperature data from the thermal simulations. The simulation successfully predicted the temperature distribution during the welding process, with peak temperatures at the weld interface matching closely with those measured in the experiments. This validation demonstrated that the thermal model could accurately simulate the friction welding process for 15CdV6 steel. The agreement between the experimental and simulation data confirmed the accuracy of the model, which could be used to optimize friction welding parameters in future applications. The experimental results also showed that the welding procedure effectively generated the necessary heat and plastic deformation for strong, reliable joints, further validating the proposed methodology for simulating and optimizing friction welding processes in high-performance materials like 15CdV6 steel. The study highlighted the utility of ANSYS APDL as a reliable tool for predicting thermal conditions and improving the quality of welds in similar industrial applications.

Table 1 Chemical composition of 15CdV6 alloy steel

Material	C	Mn	Si	Mo	V	Cr	P	S	Fe
Base	0.120	0.95	0.2	0.9	0.25	1.35	0.02	0.0015	Bal

Table 2 Physical properties of 15CdV6 alloy steel [14]

UTS (Mpa)	Yield strength (Mpa)	%Elongation (Mpa)	Hardness (HV)	Impact strength (J)
980–1250	790–930	012–023	270–360	25

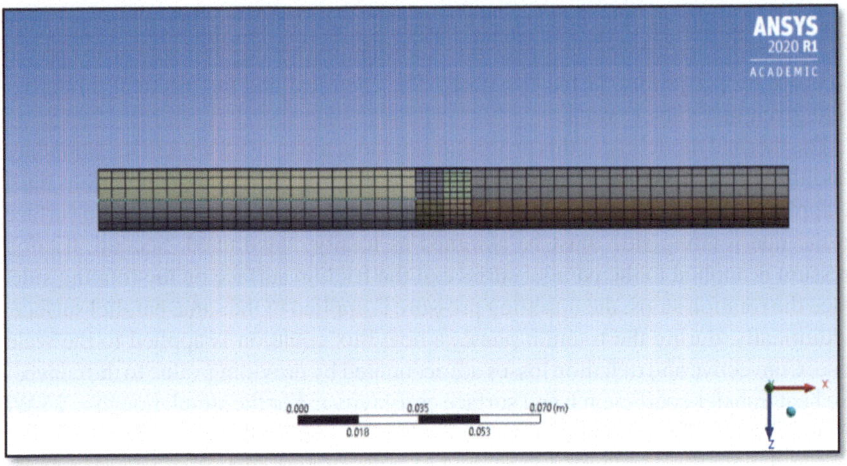

Fig. 1 Geometry and meshing

Fig. 2 Experimental view of **a** Friction welding Apparatus, **b** Before welding, **c** After welding, **d** Welded specimen with set of process parameters

3 Result and Discussion

3.1 Simulation

The thermal profiles for four different values of upsetting and friction pressures, 30, 50, 70, and 90 Mpa are Figs. 3, 4, 5, and 6, respectively.

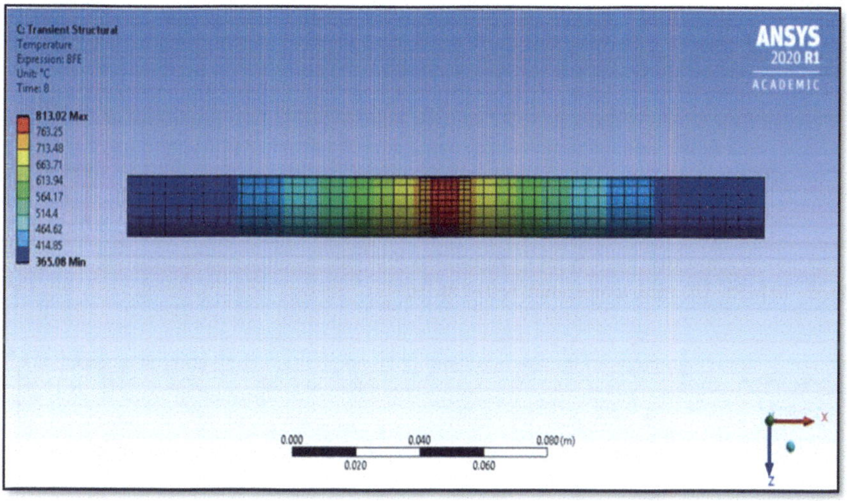

Fig. 3 30 Mpa. The Peak temperature is 813.02 ° C

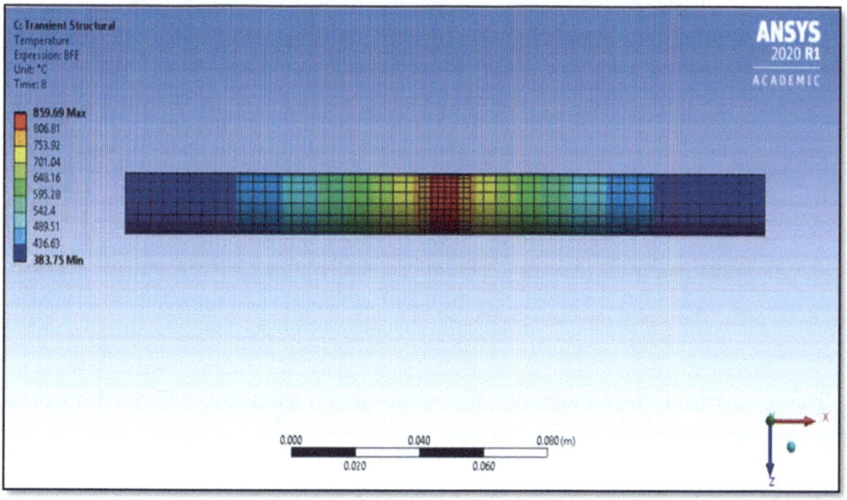

Fig. 4. 50 Mpa. The peak temperature is 859.69 ° C

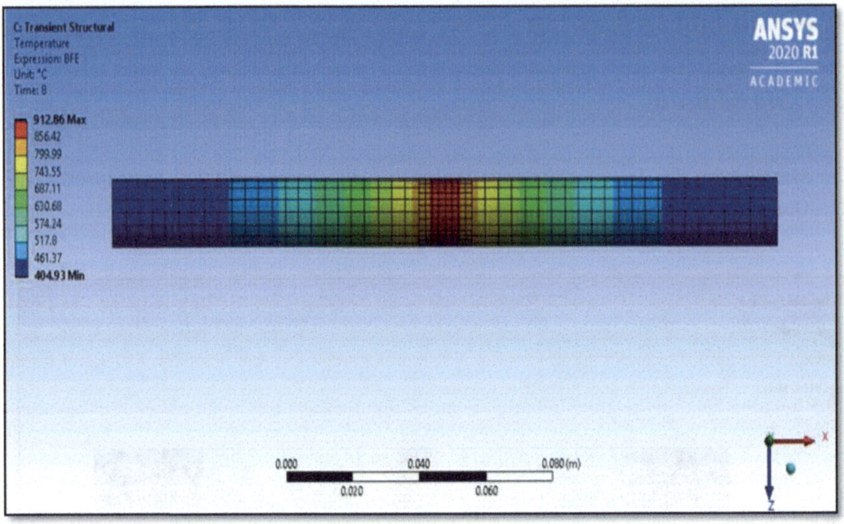

Fig. 5 70 MPa. The peak temperature is 912.86 ° C

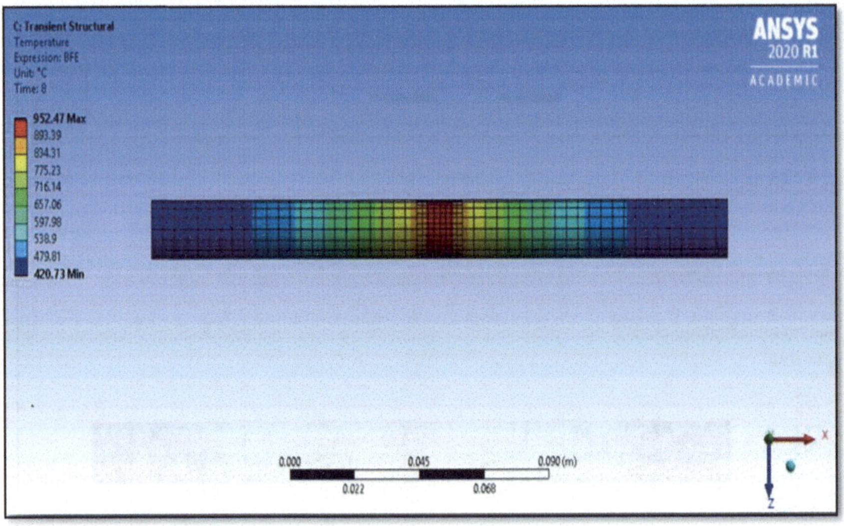

Fig. 6 90 MPa. The peak temperature is 952.47 ° C

The results from the simulation are organized and tabulated in Table 3 for clear reference and analysis.

Table 3 Readings from the simulation

Rotational speed (rpm)	Friction pressure (MPa)	Upsetting pressure (MPa)	Friction time (s)	Upsetting time (s)	Peak temperature from simulation(°c)
1500	30	30	8	8	813 0.02
1500	50	50	8	8	859.69
1500	70	70	8	8	912.86
1500	90	90	8	8	952.47

3.2 Experiment Results

The peak temperature results from the experiments can be tabulated in Table 4.

The analysis of the simulation and experimental data from) Figs. 3, 4, 5 and 6 reveals a clear and consistent correlation between friction pressure, upsetting pressure, and peak temperature during the process. The findings show that as pressure increases, the peak temperature also rises [13]. For example, at 30 MPa, the modeling temperature was 813.02 °C, while the experimental temperature was 826 °C. At 50 MPa, the modeling temperature was 859.69 °C, compared to 857 °C experimentally. At 70 MPa, the modeling temperature reached 912.86 °C, while the experimental temperature was 922 °C, and at 90 MPa, the modeling temperature peaked at 952.47 °C, closely aligning with the experimental temperature of 979 °C. This trend confirms the relationship between pressure and temperature, as both the simulation and experimental results show that increasing friction and upsetting pressures leads to higher peak temperatures. The simulation results closely match the experimental data, suggesting that the model accurately predicts the thermal behavior under varying pressure conditions. While minor discrepancies between the two sets of results can be attributed to factors such as material properties, environmental conditions, or measurement uncertainties, these differences do not significantly alter the overall trend. The results emphasize that higher pressures increase friction and deformation, leading to elevated temperatures. This relationship is crucial for understanding thermal dynamics in processes that involve high-pressure conditions, such as forging, extrusion, and other manufacturing techniques. The study confirms that both simulation and experimental methods are effective tools for analyzing the thermal behavior

Table 4 Readings from experiment

Rotational speed (rpm)	Friction pressure (MPa)	Upsetting pressure (MPa)	Friction time (s)	Upsetting time (s)	Peak temperature(°c)
1500	30	30	8	8	826
1500	50	50	8	8	857
1500	70	70	8	8	922
1500	90	90	8	8	979

of materials under pressure. The consistent rise in temperature with increasing pressure highlights the importance of controlling pressure parameters in industrial applications to manage thermal effects. These findings provide valuable insights for optimizing process parameters to achieve desired thermal conditions, ensuring material integrity, and improving process efficiency. Key findings include the direct correlation between friction/upsetting pressure and peak temperature, validated by both simulation and experimental data. The practical implications emphasize that understanding this relationship is critical for industries relying on high-pressure processes, as it helps predict and control thermal effects, thus preventing material degradation and enhancing product quality. Future research could further investigate the effects of different materials, lubrication, and cooling mechanisms on the pressure–temperature relationship, refining the model and expanding its applicability.

The simulation results show that the highest peak temperature of 952.47 °C occurs at 90 MPa, while the experimental data indicates a peak temperature of 979 °C at the same pressure. Both the simulation and experimental results clearly demonstrate that as the friction pressure and upsetting pressure increase, the peak temperature also rises. This correlation highlights the impact of pressure variations on temperature behavior, with higher pressures leading to elevated peak temperatures during the process [16]. Thus, both methods align in showing that increased pressure contributes to higher thermal levels in the system.

3.3 Comparison between Simulated and Experimental Results

The peak temperature in the weld region increases as both upsetting and friction pressure are raised. The simulated results exhibit an error margin of less than 5%, which indicates that the numerical model is highly effective in accurately predicting the peak temperature in the weld with minimal error. This suggests a high degree of reliability in the model's performance. A graphical plot and the corresponding percentile error, which further illustrate the accuracy of the simulation, are provided in the Fig. 7. These findings confirm the model's precision in forecasting the temperature distribution within the weld area under varying pressure conditions.

4 Conclusion

In this study, thermal modeling of rotary friction welding of 15CdV6 alloy steel is made on ANSYS APDL. The peak temperature obtained from the simulation is noted and compared against the experimentally obtained peak temperature. The following conclusions are derived from the present study:

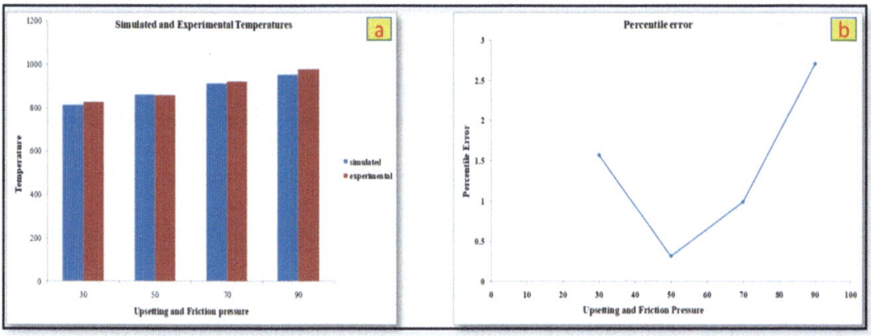

Fig. 7 **a** Simulated and experimental temperatures, **b** Percentile error

- The peak temperature increases for increase in friction pressure. This is because there exist a direct correlation between the frictional heat generation and frictional pressure, thus leads to the increase in peak temperature.
- The error in the peak temperature values obtained from the simulation and experiment is may be due to the frictional coefficient between the contact surfaces is assumed to be independent of the process parameters. However in the actual case this is not so. The distribution of the frictional pressure on the contact surfaces is considered to be uniform, whereas in the actual process, this distribution is non-uniform.

Acknowledgements The authors sincerely thank the Mechanical Engineering Department, College of engineering, Trivandrum, and Annamalai University, Chidambaram, for their invaluable support during this research.

Author Contributions

- Krishnalal M S: Methodology, validation, interpretation of data for the work, writing—Original Draft.
- Mahendra Raj E S: Experimentation, technical discussion.
- P Anchana: Experimentation, technical discussion.
- PM Ajith: Conceptualization, methodology, investigation.
- Sunilkumar Dhasan: Conceptualization, methodology, investigation.

Data Availability All data generated and analyzed during this study are available in the article for readers to access and use.

Declarations

Conflict of Interest The authors declare no competing financial interests or personal relationships influencing the work reported in this paper.

Ethical Approval No experiments involved human tissues in this article.

References

1. Weman K (2011) Welding processes handbook. Elsevier
2. Mir FA et al (2022) Friction based solid state welding–a review. Mater Today Proc 62:55–62
3. Al-Qenaei AAY (2016) Fusion welding techniques. Int J Eng Res Appl 6(3):78–83
4. Shete N, Deokar SU (2017) A review paper on rotary friction welding. In: International conference on ideas, impact and innovation in mechanical engineering. Vol 5 No 6
5. Dawood AB et al (2017) Thermal model of rotary friction welding for similar and dissimilar metals. Meta 7(6):224
6. Gaikwad VT et al (2024) On the role of upset pressure during friction welding of IN713LC and AISI 4140. Mater Sci Technol. https://doi.org/10.1177/02670836241273483
7. Anchana P, Ajith PM (2022) Multiobjective optimization of friction welding of 15CDV6 alloy steel rods using grey relational analysis in the taguchi method. In: International conference on intelligent vision and computing. Springer Nature Switzerland, Cham
8. Srinivasan L et al (2018) Application of GRA and TOPSIS optimization techniques in GTA welding of 15CDV6 aerospace material. Trans Indian Inst Metals 71:373–382
9. Jalaja K et al (2024) Investigation of crack-like defect in 15CDV6 (ESR) steel forging for aerospace applications. Eng Fail Anal 158:108040
10. Srinivasan L et al (2017) Effect of heat input, heat treatment on microstructure and mechanical properties of GTA welded aerospace material 15CDV6. J Mater Res 32(7):1361–1366
11. Ramesh MVL, Srinivasa Rao P, Venkateswara Rao V (2015) Microstructure and mechanical properties of laser beam welds of 15CDV6 steel. Def Sci J 65(4):339–342
12. Ramesh MVL et al (2015) Structure–properties evaluation in laser beam welds of high strength low alloy steel. Mater Today Proc 2(4–5):2532–2537
13. Al-Nuaimi A et al (2024) The effect of periodic friction and upsetting pressure on rotary friction welding process. Al-Rafidain Eng J 29(2):65–75
14. Srinivasan L, Jakka SJ, Sathiya P (2017) Microstructure and mechanical properties of gas tungsten arc welded high strength low alloy (15CDV6) steel joints. Mater Today Proc 4(8):8874–8882
15. Childs Peter RN (2015) Temperature measurement. Mechanical engineers handbook, volume 4: energy and power:247
16. Hosford WF, Caddell RM (2011) Metal forming: mechanics and metallurgy. Cambridge university press

Efficiency Enhancement of ACZTS/ CIGSSE Solar Cell

Raushan Kumar, Alok Priyadarshi, Shivam Kumar, and Alisha Priya

Abstract This study focuses on the efficiency enhancement of hybrid ACZTS/ CIGSSe solar cells, combining the advantages of earth-abundant materials and established high-efficiency frameworks. ACZTS, a kesterite-based material, is known for its cost-effectiveness and environmental compatibility but suffers from efficiency limitations due to intrinsic defects and suboptimal carrier transport. On the other hand, CIGSSe exhibits superior optoelectronic properties but faces challenges in scalability and material costs. We present a novel tandem configuration integrating ACZTS as the top cell and CIGSSe as the bottom cell, leveraging their complementary bandgap alignment to maximize light absorption across the solar spectrum. By employing advanced interface engineering techniques, including surface passivation and optimized ZnO: B buffer layer deposition, we significantly reduced recombination losses at the heterojunction interfaces. Additionally, nanostructured back reflectors and anti-reflective coatings were introduced to enhance light trapping and photon recycling. The AFORS-HET simulation results demonstrate a marked improvement in power conversion efficiency (PCE), achieving over 21.95% under standard test conditions. This improvement is attributed to enhanced charge carrier dynamics, minimized defect densities, and improved optical management. The proposed hybrid

R. Kumar (✉)
Department of Electronics and Communication Engineering, School of Engineering, SR University, Warangal, Telangana, India
e-mail: 2019rsec001@nitjsr.ac.in

A. Priyadarshi
Department of Electrical Engineering, National Institute of Technology, Jamshedpur, India
e-mail: apriyadarshi.ee@nitjsr.ac.in

S. Kumar
Department of Electronics and Communication Engineering, National Institute of Technology, Jamshedpur, India
e-mail: skgupta940@gmail.com

A. Priya
Department of Electronics and Communication Engineering, JSPM University, Pune, Maharashtra, India
e-mail: ap.secs@jspmuni.ac.in

A. Biswas et al. (eds.), *Recent Advancements in Artificial Intelligence and Optimization*, Lecture Notes in Mechanical Engineering,
https://doi.org/10.1007/978-981-95-9268-5_16

solar cell design not only addresses the efficiency bottlenecks of individual materials but also offers a scalable and cost-competitive solution for next-generation photovoltaics. Our findings pave the way for advancing sustainable solar technologies with a focus on performance, affordability, and environmental compatibility.

Keywords ACZTS · CIGSSe · ZnMgO: Al · Buffer layer · ZnO: B

1 Introduction

As global energy consumption continues to rise, the transition from fossil fuels to sustainable energy sources has become an urgent priority. Solar photovoltaics (PV) is a leading technology in this transition, offering a clean, renewable, and scalable energy solution. However, the extensive acceptance of solar energy differs on the development of high-efficiency, cost-effective, and environmentally friendly solar cells. Thin-film solar cell technologies, particularly those based on ACZTS (Ag-doped Cu2ZnSnS4) and CIGSSe, have emerged as promising candidates for meeting these requirements. This study focuses on the efficiency enhancement of a hybrid ZnMgO: Al/ZnO: B/ACZTS/CIGSSe solar cell, a multi-layered device designed to optimize light absorption, charge transport, and overall performance [1]. Thin-film solar cells are attractive due to their reduced material usage, lightweight construction, and flexibility compared to traditional crystalline silicon cells. Among the thin-film technologies, kesterite-based ACZTS and chalcopyrite-based CIGSSe materials have garnered significant attention. ACZTS is composed of earth profuse, harmless elements such as copper, zinc, tin, and sulfur, making it an environmentally sustainable option. Its low production cost further enhances its appeal. However, ACZTS faces efficiency limitations due to intrinsic defects, poor carrier mobility, and high recombination losses. In contrast, CIGSSe has demonstrated high efficiencies exceeding 20% in single-junction configurations, primarily due to its superior optoelectronic properties and tunable bandgap. However, its reliance on scarce elements like indium and gallium growths production costs and advances concerns about long-term scalability [2]. Combining ACZTS and CIGSSe in a tandem configuration can address the individual limitations of these materials, creating a synergy that enhances overall device performance. The design of the ZnMgO: Al/ZnO: B/ACZTS/CIGSSe solar cell incorporates several advanced strategies to maximize efficiency. **Interface Engineering**: Advanced passivation techniques reduce defect states and improve band alignment at critical interfaces, minimizing recombination losses and enhancing charge transfer. **Bandgap Optimization**: The complementary bandgap alignment of ACZTS and CIGSSe enables efficient absorption of a broader range of wavelengths, increasing the short-circuit current density. **Material Doping:** Strategic doping of the ZnMgO, ZnO, and ACZTS layers enhances their electronic and optical properties, improving conductivity and light transmission. **Light Management**: Anti-reflective coatings and nanostructured surfaces maximize photon capture, while back reflectors ensure efficient light trapping within the device.

The proposed ZnMgO: Al/ZnO: B/ACZTS/CIGSSe solar cell addresses critical challenges in thin-film photovoltaics, including efficiency limitations, cost constraints, and environmental impact. By combining the complementary strengths of ACZTS and CIGSSe in a tandem configuration, this design achieves a balance between performance and sustainability. The integration of ZnMgO: Al and ZnO: B layers further enhances optical and electrical properties, ensuring efficient carrier dynamics and reduced losses.

2 Proposed Device Model

The hybrid solar cell under investigation incorporates a multi-layered architecture; each layer is designed to perform a specific function in maximizing efficiency: ZnMgO: Al (transparent conducting oxide). As the topmost layer, ZnMgO: Al serves as the transparent conducting oxide (TCO). Magnesium doping increases the optical bandgap of ZnO, improving light transmission and reducing parasitic absorption. Aluminum doping enhances electrical conductivity, minimizing resistive losses, and ensuring efficient carrier collection. Boron-doped ZnO acts as a shield film between the TCO and the ACZTS absorber film [3]. This buffer layer plays a critical role in reducing lattice mismatches, thereby lowering defect densities and recombination losses at the interface. It also facilitates smooth charge transfer between layers, contributing to higher fill factors. The ACZTS layer, doped with silver to improve its structural and electronic properties, captures high-energy photons from the solar spectrum. Silver doping reduces antistite defects, enhances carrier mobility, and improves the V_{OC} of the device. The CIGSSe layer absorbs lower-energy photons, complementing the ACZTS layer to ensure comprehensive utilization of the solar spectrum. This layer benefits from its high efficiency and excellent charge collection capabilities, which significantly contribute to the device's overall performance. Each layer parameter (listed in Table 1) is inserted into the AFROS-HET simulation tool to model a proposed cell which is reflected in Fig. 1a. After simulation analysis we got good results in terms of band diagram, EQE, and J–V curvature of proposed device as presented in Fig. 1b–d.

3 Results and Discussion

3.1 Impact of N_a Density on Recommended Device

Acceptor doping concentrations between 10^{16} and 10^{18} cm^{-3} can significantly affect the performance of device, influencing electrical, optical, and structural properties. At moderate acceptor concentrations of 10^{17} cm^{-3}, hole concentration increases, improving electrical conductivity and the formation of the p–n junction. This leads to

Table 1 Parameters of solar cell [1–4]

Parameters	ZnMgO: Al	ZnO: B	ACZTSSe	CIGSSe
Thickness (nm)	100	50	600	100
E_g (eV)	3.8	3.31	1.3	1.4
χ (eV)	4.5	4.5	4.5	4.371
ε_r	10	9	13.6	13.5
N_C (cm^{-3})	2.2×10^{18}	2×10^{18}	2.2×10^{17}	2×10^{18}
N_V (cm^{-3})	1.8×10^{19}	1.8×10^{19}	1.8×10^{19}	1.8×10^{19}
μ_e (cm^2 V^{-1} s^{-1})	100	100	100	100
Mh (cm2 V-1 s-1)	10	20	25	20
N_d (cm^{-3})	10^{20}	10^{20}	0	0
N_A (cm^{-3})	0	0	10^{17}	10^{17}
V_e (cm^2s^{-1})	10^7	10^7	10^7	10^7
V_h (cm^2s^{-1})	10^7	10^7	10^7	10^7
Rae (cm6s-1)	3 × 10–32	1.12×10^{-30}	10–29	10–28
Rah (cm6s-1)	2 × 10–32	1.12×10^{-30}	10–29	10–28
Rbb (cm3s-1)	1.1×10^{-11}	1.2×10^{-11}	5×10^{-9}	8×10^{11}

enhanced charge transport and better current generation. However, at higher doping concentrations of 10^{18} cm^{-3}, excessive hole density can cause carrier freeze-out, reducing carrier mobility and conductivity, which can negatively affect the device's performance as shown in Fig. 2a. High acceptor doping can also elevate recombination rates, where photogenerated electron-hole pairs recombine before contributing to current, reducing the solar cell's efficiency. Additionally, minority carrier lifetime decreases with higher doping levels, limiting the time available for carriers to diffuse and reach the junction, further lowering efficiency. Bandgap narrowing occurs at higher doping levels due to impurity band formation, which affects the material's ability to absorb light effectively. The increase in hole concentration shifts the Fermi level toward the valence band, altering the E_{bi} at the p–n junction, potentially impacting charge separation and collection efficiency. Excessive doping can also degrade junction formation and long-term device stability by introducing defects or increasing diffusion.

Similarly, the acceptor doping concentration of 10^{16} to 10^{18} cm^{-3} significantly affects the quantum efficiency (QE) of a solar cell by influencing carrier collection, recombination rates, and optical properties. Figure 2b displays the QE versus wavelength curve at different N_a concentrations. At 10^{17} cm^{-3} doping levels the junction's electric field is optimized, ensuring efficient separation and collection of photogenerated carriers, leading to high QE across various wavelengths. However, higher doping levels of 10^{18} cm^{-3} reduce the depletion region width, limiting the collection of carriers generated deeper in the bulk. This negatively impacts QE for longer wavelengths that penetrate further.

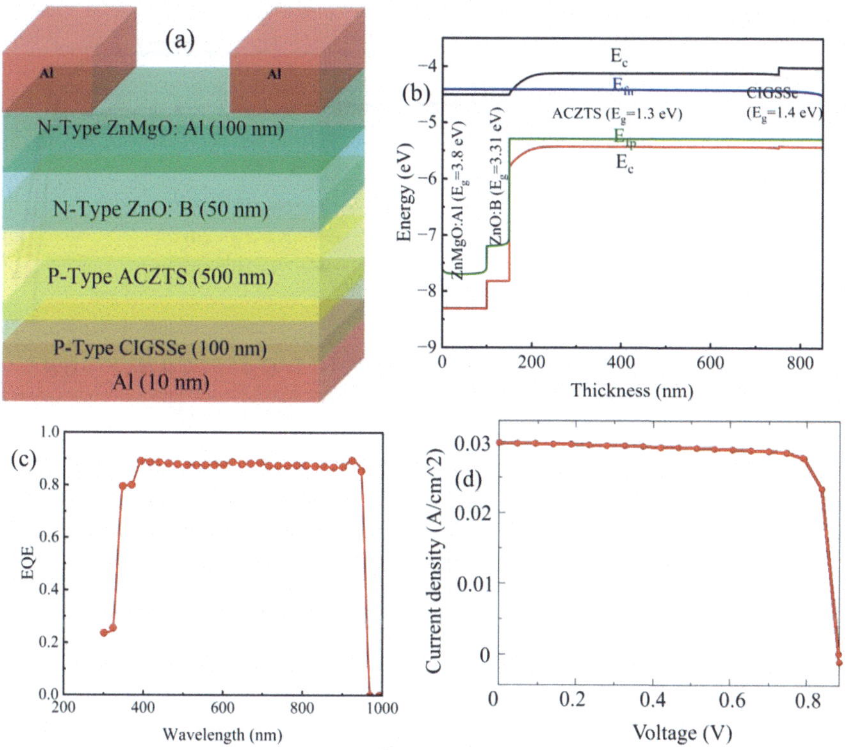

Fig. 1 Proposed cell **a** device structure, **b** band diagram, **c** EQE, and **d** J–V curve

3.2 *Effects of Temperature on Proposed Device*

The performance of device is strongly influenced by temperature variations from 200 to 500 K, affecting carrier behavior, recombination, and material properties. At higher temperatures above 300 K, intrinsic carrier concentration increases due to enhanced thermal excitation. While this boosts conductivity, it also raises recombination rates, particularly Shockley-Read-Hall (SRH) and Auger recombination, reducing the efficiency of charge collection. Temperature variations (200–500 K) impact carrier mobility, recombination rates, and bandgap energy, leading to efficiency fluctuations. In high-temperature regions, increased thermal excitation enhances recombination, reducing Voc and overall efficiency. Proper thermal management, material engineering, and tandem design help mitigate these effects, ensuring stable performance under real-world conditions. Additionally, phonon scattering increases with temperature, reducing carrier mobility and degrading charge transport. Rising temperatures from 200 to 500 K led to bandgap narrowing, which shifts the absorption spectrum to longer wavelengths. This can slightly improve absorption in the infrared range but

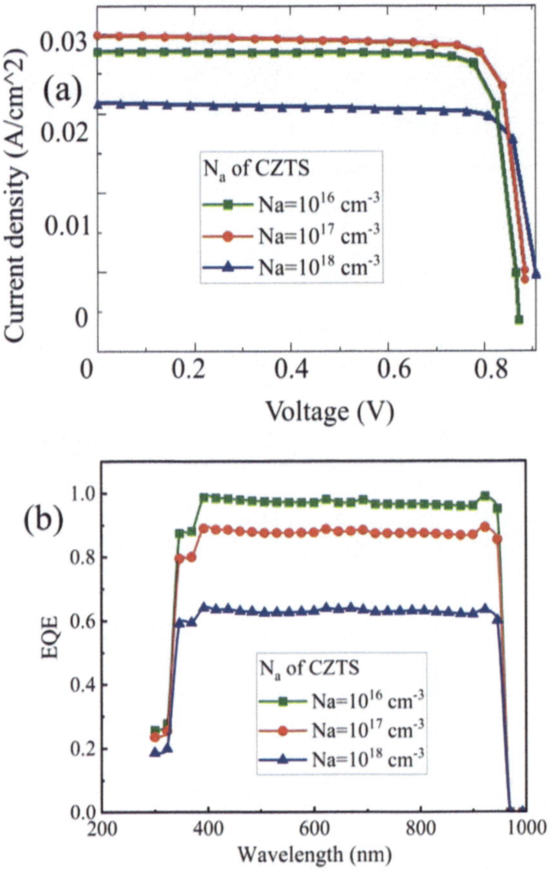

Fig. 2 Effect of N_a density with change in **a** J–V curve and **b** EQE of cell

simultaneously lowers the V_{OC}. The reduced V_{OC} directly impacts the overall efficiency of the cell as illustrated in Fig. 3. At 300 K device temperature the proposed device shows better performance.

3.3 Comparison

Table 2 displays the comparison to the proposed solar cell and previously reported work; the proposed device displays good performance in terms of efficiency.

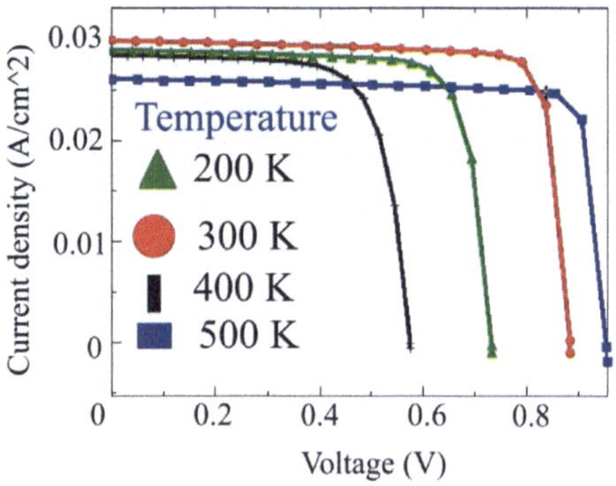

Fig. 3 Change in J–V curve at different temperatures of proposed cell

Table 2 Comparison of solar cell devices

Device	Device thickness	Efficiency (%)	Reference
ZnO:Al/i-ZnO/ZnMgO/CZTS/CZTSSe	3550 nm	17.50	[4]
ZnO/CdS/CZTSSe/Zn_3P_2/Mo	2160 nm	19.70	[5]
ZnO/CdS/CZTSSe/Mo	2070 nm	20.16	[6]
ZnMgO: Al/ZnO: B/ACZTS/CIGSSe	**750 nm**	**21.95**	**Proposed**

4 Conclusion

Enhancing the efficiency of ACZTS/CIGSSe solar cells involves optimizing material properties, interface quality, and device architecture. The ACZTS reduces defect density and enhances carrier mobility, mitigating recombination and improving charge transport. Meanwhile, precise control of the CIGSSe absorber composition ensures an ideal bandgap gradient for improved light absorption and higher V_{OC}. Interface engineering plays a critical role, with optimized band alignment at the ACZTS/CIGSSe junction and improved buffer layers reducing recombination and enhancing carrier extraction. Structural improvements, including anti-reflective coatings and light-trapping designs, maximize optical absorption, while tailored doping concentrations optimize electric fields for efficient charge separation. By integrating these advancements, ACZTS/CIGSSe solar cells can achieve higher efficiency (21.95%), stability, and cost-effectiveness, making them a promising solution for scalable and sustainable photovoltaic cell.

References

1. Kumar R, Kumar A (2022) Efficiency improvement of ZnMgO/CIGSSe heterojunction solar cell by using double graded Cu_2O ER-HTL and $CeMgO_2$ HR-ETL materials. Opt Mater 131:112697
2. Kumar R, Kumar A (2022) Performance enhancement of ZnMgO: Al/ZnMgO/CIGSSe solar cell with the combination of CZTGSe HT-ERL layer. J Electron Mater 51(1):84–103

3. Priya A, Prakash A, Singh SN, Kumar R (2022) Influence of $SnMnO_2$ window layer on enhancing the performance of CIGSSe thin-film solar cell. Optik 268:169662
4. Mohammadnejad S, Bahnamiri ZM, Maklavani SE (2020) Enhancement of the performance of kesterite thin-film solar cells using dual absorber and ZnMgO buffer layers. Superlattice Microst 144:106587
5. Babu PS, Singh PK, Thakur AK, Dwivedi DK (2021) Optimization of photovoltaic solar cell performance via the earth abundant Zn_3P_2 back surface field. Optik 229:166235
6. Sadanand S, Dwivedi DK (2020) Theoretical investigation to enhance the performance of CZTSSe solar cell via back surface field. In: AIP conference proceedings, vol 2220, (1). AIP Publishing

Computational Modeling of Friction Stir Processing for DP780 Steel

Amos Thounaojam, Yogesh Singh, and Sunilkumar Dhasan

Abstract Friction stir processing is a complex multiphysics phenomenon involving severe plastic deformation, material flow, heat transfer, and microstructure evolution, making its modeling highly challenging. Experimental approaches to optimize processing parameters are often costly and time-intensive. This study develops a computationally efficient model using COMSOL Multiphysics v6.2 to simulate the friction stir processing of DP780 steel, integrating JMatPro to provide alloy design and material properties. Key phase transitions observed include austenite formation around 1200 °C and ferrite emergence between ~800 °C and ~ 600 °C, with carbides such as $M_{23}C_6$, M_7C_3, and cementite present. Critical transition temperatures are Ac_1 at 750 °C and Ac_3 at 900 °C, while the peak temperature (~1490 °C) remains localized in the processed zone. The results demonstrate finite element analysis as a powerful tool for optimizing processing parameters, offering insights into temperature distribution, thermal history, and material flow, thus minimizing experimental efforts and costs.

Keywords Friction stir processing · FEA · Dual phase steel · Jmat pro

1 Introduction

Friction stir processing (FSP) is a mechanical deformation technique designed to enhance the surface and microstructure of materials, leveraging principles like friction stir welding (FSW) [1–6]. While FSW focuses on material joining, FSP aims primarily at refining surface properties and microstructures. The process comprises three main stages: plunging, dwelling, and stirring, employing a non-consumable, intricately designed tool that rotates at high speed to soften and mix the workpiece material without melting it [2–6]. The tool's design and profile are pivotal, influencing heat generation, material flow, and mechanical properties [7–11]. Tool geometry and

A. Thounaojam · Y. Singh · S. Dhasan (✉)
Department of Mechanical Engineering, National Institute of Technology, Silchar, Assam, India
e-mail: sunilkumar@mech.nits.ac.in

A. Biswas et al. (eds.), *Recent Advancements in Artificial Intelligence and Optimization*, Lecture Notes in Mechanical Engineering,
https://doi.org/10.1007/978-981-95-9268-5_17

rotational speed adjustments significantly affect microstructure refinement, homogenization, and corrosion resistance enhancement [11–13]. FSP is widely adopted in industries such as aerospace, automotive, and shipbuilding, offering high-quality surface modifications without melting the base material. Extensive experimental research has explored the effects of FSP on engineering alloys like aluminum, steel, magnesium, and titanium [14–19]. However, computational modeling and simulation of FSP, particularly for material modification, remain underdeveloped. The process entails complex multiphysics phenomena, including severe plastic deformation, material flow, heat transfer, and microstructure evolution, posing significant modeling challenges [20, 21]. Asymmetric conditions in heat distribution, stress, and plastic deformation, driven by variations in the tool's translational and rotational speeds, impact the final microstructure and mechanical properties [9, 13]. While experimental methods dominate FSP research, there is growing interest in computational models to reduce experimental efforts, minimize costs, and deepen understanding of process mechanics [22, 23]. This study aims to develop a computationally efficient 3D transient, non-isothermal CFD model with a moving coordinate system using COMSOL Multiphysics v6.2. This research seeks to develop a clear understanding of the FSP of DP780 steel. This method delivers a detailed prediction of temperature distribution, thermal history, and temperature profiles. These insights help to understand how process parameters influence key factors such as temperature, material flow, and strain rate.

2 Materials and Methods

This study employs JMatPro to simulate material properties for input into COMSOL Multiphysics v6.2 to analyze friction stir processing (FSP). The chemical composition of DP780 steel is determined via SEM with EDS (see Table 1). A numerical model is developed using a moving coordinate system fixed to the tool axis instead of a moving heat source. The plate has dimensions of 102 mm by 150 mm by 3 mm, and two infinitely long element domains are incorporated into the model (Fig. 1). A tungsten carbide (WC) tool, featuring only a shoulder that moves along x directions without inclination, with the workpiece and tool properties (thermal conductivity, heat capacity, density) vary with temperature. Key parameters include emissivity (ε=0.4), friction coefficient ($\mu = 0.7$), and heat transfer coefficients ($h_u = 9.25$ W/(m^2·K) and $h_d = 4.56$ W/(m^2·K)). Frictional heat fluxes, dependent on rotational speed and normal force (Fn), are set to zero above the melting temperature. Surface to ambient radiation, convection, and ambient temperature on the right plate boundary are also incorporated [20]. To ensure accuracy, the relevant mechanical and physical properties of DP780 steel are defined as temperature-dependent.

Eq. 1 models heat transfer in the plate, incorporating the effects of both convection and conduction.

$$\rho Cpu.\nabla T + \nabla(-k\nabla T) = Q \tag{1}$$

Table 1 Chemical compositions of DP780 steel for this study (fraction by mass %)

C	Cr	Mn	Mo	P	S	Si	Nb	Fe
0.095	0.3	1.8	0.25	0.012	0.005	0.003	0.02	Rest

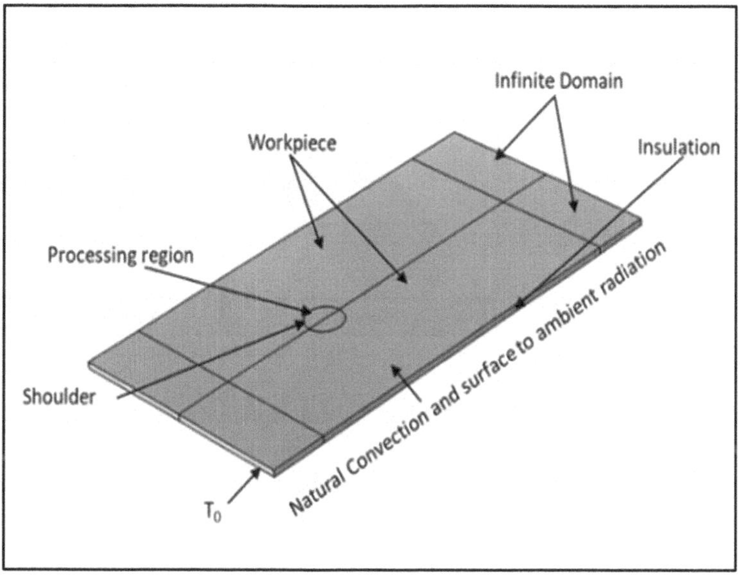

Fig. 1 Model geometry and system domain

Eq. 2 represents the heat generated by friction between the workpiece and the tool's shoulder, expressed as heat flux (W/m^2) at a distance r from the center axis [21].

$$q_{\text{shoulder}}(r, T) = \begin{cases} \left(\frac{\mu Fn}{As}\right)\omega r & \text{if } T < T_{\text{melt}} \\ 0 & \text{if } T \geq T_{\text{melt}} \end{cases}. \tag{2}$$

When the local temperature reaches or exceeds the plate's melting point, friction between the tool and the plate becomes negligible, ceasing heat generation. Eq. 3 also accounts for heat loss from natural convection and surface-to-ambient radiation.

$$\begin{aligned} q_u &= h_u(T_0 - T) + \varepsilon\sigma\left(T_{\text{amb}}^4 - T^4\right) \\ q_d &= h_d(T_0 - T) + \varepsilon\sigma\left(T_{\text{amb}}^4 - T^4\right). \end{aligned} \tag{3}$$

Flow stress property by Johnson Cook model (Eq. 4)

$$\sigma_{JC} = \left[A + B\varepsilon^n\right].\left[1 + C\ln\left(\frac{\varepsilon^{\cdot}}{\varepsilon_0^{\cdot}}\right)\right]\left[1 - \left(\frac{T - T_0}{Tm - T_0}\right)^m\right]. \tag{4}$$

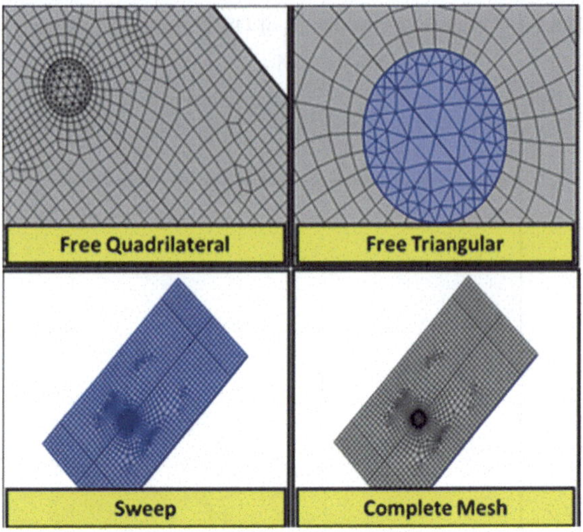

Fig. 2 Meshing system

For thermal softening of DP780, Modified Johnson Cook model (Eq. 5)

$$\begin{aligned} T &< 700°C, T_0 = 21, m = 1.8 \\ T &\geq 700°C, T_0 = 700, m = 0.0012 \times T + 1.85 \end{aligned} \tag{5}$$

As shown in Fig. 2, the meshing system is a mixture of tetrahedral, prism, quadrilateral, and triangular elements to balance computational efficiency and accuracy, with an average element quality of 0.786. Poor quality elements were minimized, especially in high-strain and thermal gradient regions.

A refined meshing strategy was applied in the stir zone with a gradual element size transition to maintain numerical stability. A mesh convergence study (60,000, 92,000, and 130,000 elements) showed that 92,000 elements provided optimal accuracy with minimal computational cost.

3 Results and Discussion

The provided material properties and parameters are employed to validate the modeling as expressed in Table 2.

The flow stress diagram (Fig. 3) reveals distinct mechanical behavior across different temperature ranges. At low temperatures (25–200 °C), stress increases significantly with strain due to strain hardening, while at higher temperatures (800–1250 °C), thermal softening dominates as dynamic recovery and recrystallization occur. In the 600–800 °C range, initial strain hardening is followed by steady-state deformation, indicating a plateau in strain hardening. The graphical presentation of

Table 2 Material properties simulated by JMatPro for the given DP780 steel composition

Parameter	Unit	DP780 steel	WC
Melting temperature, T_m	°C	1485	2873
Density, ρ	g/cm^3	7.85	15.6
Young's modulus, E	GPa	200	700
Poison ratio, v	–	0.3	0.24
Strength coefficient, K	MPa	750	–
Hardening exponent, n	–	0.25	–
Strain rate strength coefficient, C	–	0.05	–
Reference strain rate, $\varepsilon_0°$	s^{-1}	1	–
Temperature exponent, m	–	0.05	–
Thermal conductivity, k	W/m.K	50	110
Specific heat capacity, C_p	J/kg.K	470	320
Coefficient of thermal expansion, α	1/K	11.7e-6	6.5e-6

the thermal property diagram (Fig. 4) highlights critical microstructural transformations, with austenite forming around 1200 °C and ferrite emerging between 800 and 600 °C. Carbide phases such as $M_{23}C_6$, M_7C_3, and cementite are also present, influencing material strength. Key transition temperatures include Ac_1 at 750 °C and Ac_3 at 900 °C, crucial for friction stir processing (FSP) optimization. The ideal FSP temperature range of 750–900 °C ensures refined grain structures while avoiding excessive grain coarsening and phase instability (JMatPro simulation, Figs. 3 and 4).

In this configuration, the results obtained from the 2D temperature distribution profile analysis of the DP780 steels are illustrated in Fig. 5a. The maximum temperature reaches 1490 °C, slightly exceeding the melting point of DP780 steel (1480 °C) in Fig. 5b. This profile offers a deeper understanding of the temperature distribution across the plates, showing that the highest temperatures are concentrated around the welding regions, with the temperature remaining nearly uniform in the x-direction. A thin layer of DP780 melts at the interface between the shoulder and the workpiece. Colder material is introduced to the processing tool as the plate moves, while friction-induced heat is transported. The findings enhance understanding of the influence of process parameters on temperature, material flow, and strain rate.

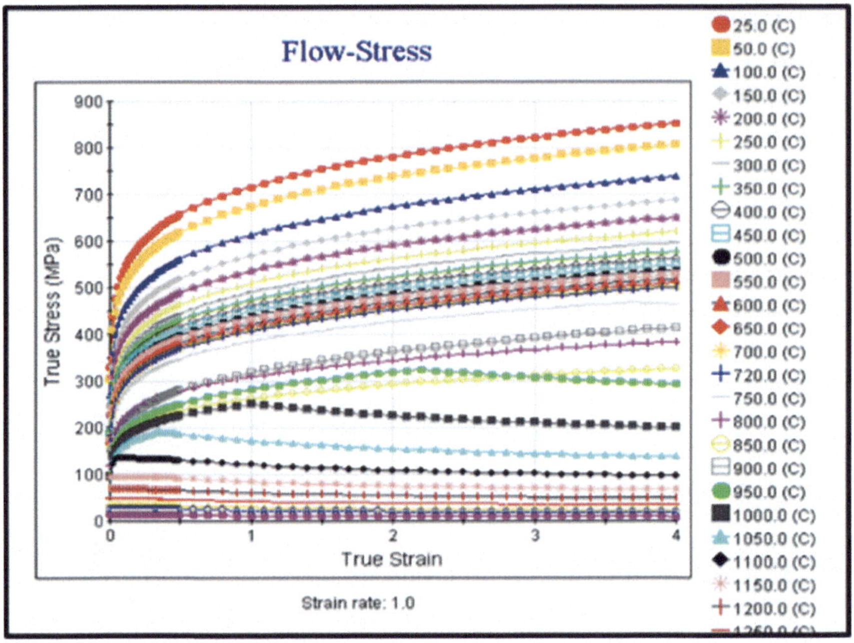

Fig. 3 Flow stress diagram

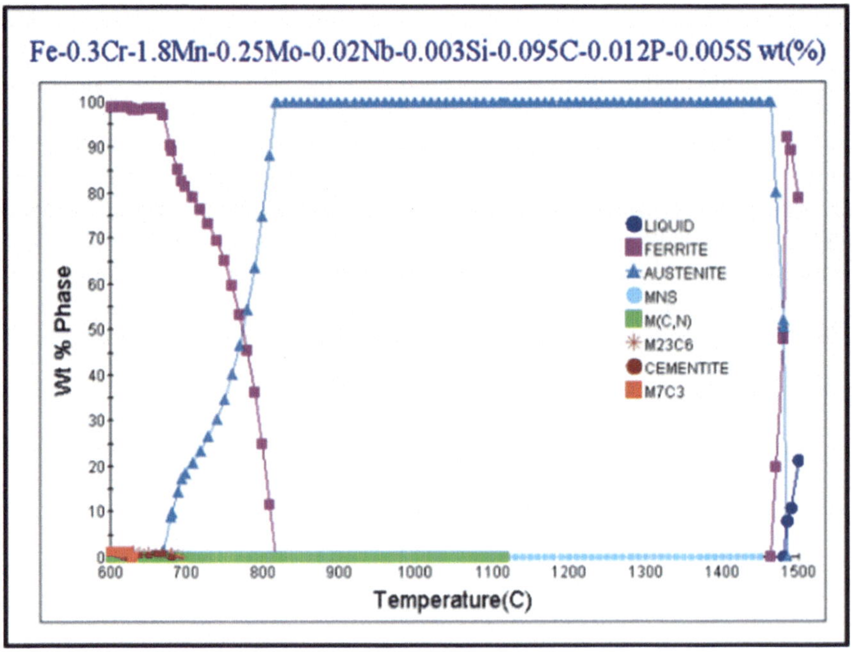

Fig. 4 Simulated phases outlined with the function of temperature

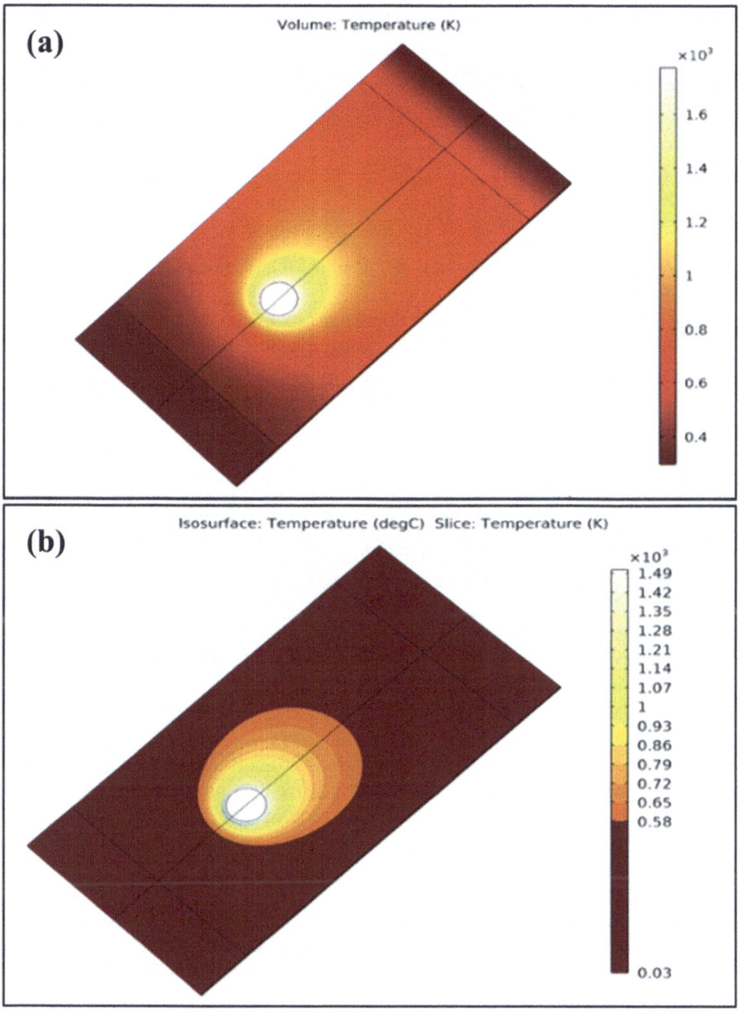

Fig. 5 **a** Isothermal contours in the domain **b** Temperature field in the DP780 plate

4 Conclusion

This study developed a 3D transient, non-isothermal CFD model using a moving coordinate system aligned with the tool axis. Key findings include a localized peak temperature of 1490 °C near the tool-workpiece contact area and critical phase transition temperatures Ac_1 (750 °C) and Ac_3 (900 °C), identified using JMatPro. These results provide valuable insights into temperature distribution, material flow, and strain rate, enhancing the understanding of how process parameters such as tool speed, feed rate, and heat input influence microstructural evolution and material performance. The findings offer practical implications for optimizing manufacturing processes, improving efficiency, and reducing costs, making them highly relevant for advanced industrial applications.

References

1. Thomas WM, Nicholas ED, Needham JC, Murch MG, Temple-Smith P, Dawes CJ (1995) Friction welding. US Patent
2. Mishra RS, Ma ZY (2005) Friction stir welding and processing. Mater Sci Eng R 50(1–2):1–78
3. Ma ZY (2008) Friction stir processing technology: a review. Metall Mater Trans A 39(3):642–658
4. Miranda R, Gandra J, Vilaça P (2013) Surface modification by friction based processes. In: Modern surface engineering treatments, pp 1–20
5. Givi MKB, Asadi P (2014) Advances in friction stir welding and processing. Woodhead Publishing Ltd, Amsterdam
6. Węglowski MS (2018) Friction stir processing, state of the art. Arch Civ Mech Eng 18(1):114–129
7. Zimmer S, Langlois L, Laye J, Bigot R (2010) Experimental investigation of the influence of the FSW plunge processing parameters on the maximum generated force and torque. Int J Adv Manuf Technol 47(1–4):201–215
8. Sunilkumar D, Muthukumaran S, Vasudevan M, Reddy GM (2019) Tool rotational speed variant response on the evolution of microstructure and its significance on mechanical properties of friction stir welded 9Cr-1Mo steel. J Mater Process Technol 2019:116536
9. Sunilkumar D, Mathew J, Muthukumaran S, Vasudevan M (2020) Friction stir welding of 2.25Cr-1Mo steel to 316LN stainless steel. Trans Indian Inst Metals 73:1689–1693
10. Khairuddin JT, Almanar IP, Abdullah J, Hussain Z (2012) Principles and thermo-mechanical model of friction stir welding. In: Kovacevic R (ed) Welding processes, pp 191–216
11. Fehrenbacher A, Duffie NA, Ferrier NJ, Pfefferkorn FE, Zinn MR (2014) Effects of tool-workpiece interface temperature on weld quality and quality improvements through temperature control in friction stir welding. Int J Adv Manuf Technol 71(1–4):165–179
12. Palani S, Karthikeyan P (2019) Optimizing tool geometry for friction stir processing of dual—phase steels: an experimental and numerical study. Int J Adv Manuf Technol
13. Kumar S, Gupta A (2018) Effect of tool geometry and processing parameters on the properties of friction stir processed DP steel. Procedia Manuf
14. Cole EG, Fehrenbacher A, Duffie NA, Zinn MR, Pfefferkorn FE, Ferrier NJ (2014) Weld temperature effects during friction stir welding of dissimilar aluminum alloys 6061-T6 and 7075-T6. Int J Adv Manuf Technol 71(1–4):643–652
15. Karthikeyan P et al (2021) Comprehensive review on the impact of friction stir processing on steels: microstructure, properties, and applications. Materials
16. Rashid MS (1981) Dual-phase steels. Annu Rev Mater Sci 11:245–267
17. Fonstein N (2017) Dual-phase steels. In: Rana R, Singh SB (eds) Automotive steels. Woodhead Publishing, Cambridge, pp 169–216
18. Liu Q, Ma QX, Chen GQ, Cao X, Zhang S, Pan J, Zhang G, Shi Q (2018) Enhanced corrosion resistance of AZ91 magnesium alloy through refinement and homogenization of surface microstructure by friction stir processing. Corros Sci 138:284–296
19. Zhang W, Ding H, Cai M, Yang W, Li J (2018) Ultra-grain refinement and enhanced low-temperature superplasticity in a friction stir-processed Ti-6Al-4V alloy. Mater Sci Eng A 727:90–96
20. Meyghani B, Awang M, Emamian SS, Mohd Nor MK, Pedapati SR (2017) A comparison of different finite element methods in the thermal analysis of friction stir welding (FSW). Metals (Basel) 7(11):450

21. Abid NH, Abu Al-Rub RK, Palazotto AN (2017) Micromechanical finite element analysis of the effects of martensite morphology on the overall mechanical behavior of dual phase steel. Int J Solids Struct 104–105:8–24
22. Song M, Kovacevic R (2003) Thermal modeling of friction stir welding in a moving coordinate system and its validation. Int J Mach Tools Manuf 43:605–615
23. Colegrove P et al (2000) 3-dimensional flow and thermal modelling of the friction stir welding process. In: proceedings of the 2nd international symposium on friction stir welding, Gothenburg, Sweden

Experimental Analysis of Machinability of ZrB_2-B_4C Using WEDM along with Phase Determination and Mechanical Properties

Ritwik R. Podder, Partha Pratim Dey, and Sunilkumar Dhasan

Abstract In this experimental study, the influence of wire electrical discharge machining (WEDM) process parameters—pulse-on time (T_{on}), pulse-off time (T_{off}), peak current (IP), and servo voltage (SV) —on the surface roughness and machining speed of the zirconium diboride-boron carbide (ZrB_2-B_4C-5%) composite was analyzed. The analysis was conducted using the response surface methodology (RSM), revealing an optimal machining speed of 5.2 mm/min at a duty factor (DF) of 0.35 and a minimum surface roughness of 1.680 μm at DF-0.2. The fabricated composite exhibited a bulk density of 5.76 ± 0.2 g/cm^3. Furthermore, the material was characterized using optical microscopy, scanning electron microscopy (SEM), and X-ray diffraction (XRD) analysis. The processed ZrB_2-B_4C-5% composite demonstrated an average hardness of 1532 ± 10 HV.

Keywords EDM machining · ZrB_2-B_4C · Composite · RSM · SEM · XRD

1 Introduction

Ceramic is an inorganic, non-metallic, solid material composed of metal, non-metal, or metalloid atoms held together primarily by ions and covalent bonds [1]. Advanced ceramic materials include silicon carbide, tungsten carbide, zirconium diboride, and aluminum oxide. Advanced ceramics are used in a wide range of applications, such as in medical instruments, electronics, high-temperature applications, aerospace, etc. Ceramics have a high melting point, chemical resistance, high hardness, low ductility, high thermal resistance, poor electrical conductivity, and are brittle in nature [2]. Zirconium diboride (ZrB_2) has a hexagonal crystal structure and is highly covalent

R. R. Podder · S. Dhasan (✉)
Department of Mechanical Engineering, National Institute of Technology, Silchar, Assam, India
e-mail: sunilkumar@mech.nits.ac.in

P. P. Dey
Department of Mechanical Engineering, Indian Institute of Engineering Science and Technology, Shibpur, India

A. Biswas et al. (eds.), *Recent Advancements in Artificial Intelligence and Optimization*, Lecture Notes in Mechanical Engineering,
https://doi.org/10.1007/978-981-95-9268-5_18

refractory ceramic material. It is a unique material with comparatively high electrical conductivities, extremely high melting point, low relative density, and high-temperature strength, making it a promising material for aerospace applications, including hypersonic flight and rocket propulsion systems [3]. ZrB_2 and its composite are extremely hard, which makes it physically and economically challenging to cut using conventional machining techniques. ZrB_2 is a conductive ceramic, which makes it possible to process using a wire-cut electro-discharge machine (WEDM) although it works on a very narrow range of process parameters.

Although many research published works are available on UHTC composites but detailed research on ZrB_2-B_4C is very limited along with this it has some unique characteristics. The impact of four distinct machining parameters, (T_{on}), (T_{off}), (IP), and servo feed (SF) on ZrB_2 with WEDM, was investigated and evaluated using a fuzzy-based multi-response performance characteristics index (MPCI) method. This approach assessed the machine's responses in terms of surface roughness and machining speed for sintered boron carbide. The study demonstrated that transforming complex multi-response analyses into a single response value is an efficient technique [4]. Additionally, the machining of ZrB_2-SiC composites using the (WEDM) technique was studied, focusing on varying SiC content (5, 15, 20, and 25 wt%). It was observed that as the SiC content increased from 5 wt.% to 20 wt.%, the material removal rate decreased, while surface roughness improved [5]. Further research on ZrB_2-SiC-based ceramic composites, specifically short carbon fiber-toughened composites (Csf/ZrB_2-SiC), examined their microstructure and mechanical properties. The addition of short fibers increased fracture toughness, 4.25 MPa·$m^{1/2}$ for ZrB_2-SiC to 6.56 MPa·$m^{1/2}$ for Csf/ZrB_2-SiC. However, the flexural strength of Csf/ZrB_2-SiC decreased by 11% to 445 MPa, attributed to the formation of a graphitization transition layer between the fibers and the matrix [6]. In another study, SiC whisker-reinforced ZrB_2 ceramic matrix composites containing 20 vol.% SiC whiskers were fabricated via hot-pressing at 1800 °C under a uniaxial load of 40 MPa. The introduction of SiC whiskers significantly improved mechanical properties, particularly fracture toughness [7]. The primary aim of this study is to evaluate various types of statistical and mathematical models developed using RSM techniques to predict the characteristic relation of WEDM input parameters for machining of ZrB_2-B_4C-5% and formation of a regression equation. The research also encompasses an analysis of mechanical properties, along with microstructural and phase characterization using (SEM) and (XRD).

2 Experimental Procedure

In this present experiment, ZrB_2-B_4C–5% wt specimens in disk form (Ø35mm X 5 mm depth) were used in the present study. Powders of boron carbide (B_4C Grade HD15) and zirconium diboride (ZrB_2-Grade B) were purchased from H.C. Starck, Germany. According to company static, the impurities in ZrB_2 powder are 1.5% oxygen, 0.2% carbon, 0.25% nitrogen, 0.25% Fe, and 0.1% Hf, and the impurities in B_4C are 1.8% oxygen, 0.7% nitrogen, 0.15% Si, 0.1% Fe, and 0.05% Al. The processed ZrB_2 and B_4C powders are dried at 150 °C for 2 hr. in an induction muffle furnace and followed by a spark plasma sintering furnace (SPS) (HP D 25, FCT System GmbH, Germany) at 2100 °C under the imposed pressure of 50 MPa in argon atmosphere [8]. After sintering, the specimen may have adhered contamination and graphic flex, which must be removed before any operation. The sintered specimens were metallographically polished down to 1 µm using the series grade of silicon carbide abrasive paper and followed by diamond paste. The mirror-finished polished specimens are cleaned with ultrasonic cleaning for 1 hour. A well-polished and cleaned specimen was taken for the microstructural analysis. An optical microscope (AXIO-vert. A1, Zeiss, Germany) was used to examine the microstructure of ZrB_2-B_4C. For crystallographic research utilizing an optical microscope, proper surface polishing and etching are essential considerations. It also removes a thin layer of a highly deformed layer introduced during grinding and polishing [9]. For ZrB_2-B_4C a solution of HNO_3 (20%), HF (20%), and distilled water in a ratio of 2:1:2 has been used as an etching regent.

A field emission scanning electron microscope (FESEM), Hariba EMAXs 3400 N, JEOL, USA (Fig. 1c), was used to analyze the microstructure of ZrB_2-B_4C composites. XRD was employed in the study to determine phases in the base metal at room temperature. In the XRD experiment, a Cu target source was employed with a scanning range of 10 to 99 and a step size of 0.01, and operating voltage and current of 40 kV and 30 mA, respectively. Vickers micro-hardness tester (Duramin-40 M1, Struers, Cleveland, United States) was used to perform a micro-hardness measurement with a load of 300 gmf and a dwell duration of 10 sec. Bulk density was determined using the standard Archimedes' principle apparatus with double-distilled water as the medium. The theoretical densities of each component were calculated using the ZrB_2 and B_4C rule-of-mixtures theories. For machinability analysis, a five-axis WEDM (Electronica Sprintcut Mark-I, 734, India) (Fig. 1c) was employed for the evolution of machining performance. The diffused brass wire (Grade H) of diameter (Ø) 0.25 mm was used as the electrode tool at a temperature of 25.5 °C. The experiment's design was based on an L16 orthogonal array, and it took into account four process variables T_{on}, T_{off}, IP, and SV. These parameters were chosen because it has the most significant effect on machining performance, which has been considered after various trial runs. Throughout the experiment, all other machining parameters such as water pressure, water flow rate, wire feed, and wire tension were kept constant. Significant output parameters such as average surface roughness (R_a) and machining speed were used to assess the machining performance. A stopwatch tool was used to record the machining time from which the material removal rate (mm^3/min) was calculated. A contact-type profilometer (Mitutoyo SJ-210, USA) was used to evaluate the surface roughness of the specimen. Table 2 lists the experimental findings and data that were gathered while the specimen was being machined. WEDM machining speed has been determined using Eq. (1). Duty factor (DF) was considered in this analysis. It is the ratio of pulse-on time to pulse-off time, as expressed in Eq. (2).

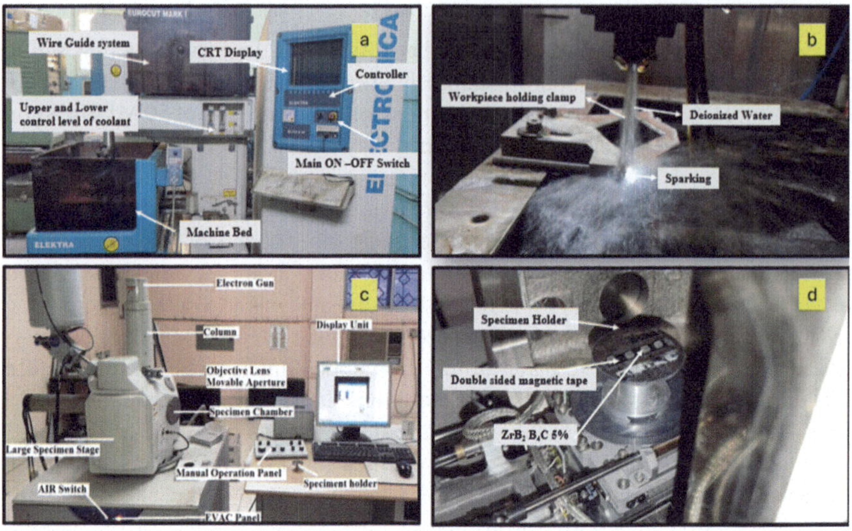

Fig. 1 **a** Electronica Eurocut mark I 734 (WEDM); **b** Machining with WEDM; **c** Hariba EMAXs3400N, JEOL, USA; **d** SEM analysis

$$S = C \times L\left(mm^2 \times mm^{-1}\right) \tag{1}$$

$$DF = \left(T_{on} - T_{off}\right) \times T_{on} - 1 \tag{2}$$

where S represents the machining speed, C denotes the cutting rate in mm/min, L is the material thickness in mm, DF stands for the duty factor, T_{on} is the pulse-on time, and T_{off} is the pulse-off time.

3 Results and Discussion

3.1 *Microstructural Analysis*

The optical image of the composite is shown in Fig. 2, and the WEDM machined surface of runs 2, 5, 12, and 15 (Table 1) with the same SV and IP but variable DF was inspected with a scanning electron microscope (SEM) (Fig. 1d). Small circular etch pits of 1–2 μm width, the hexagonal crystal structure of ZrB_2 with approximate sizes ranging from 6–10 μm, and irregular crystal alignment are observed in the optical microscope picture (Fig. 2) of the sample. The deeply etched zone (Fig. 2) cannot be clearly differentiated due to the poor resolving power of the optical microscope.

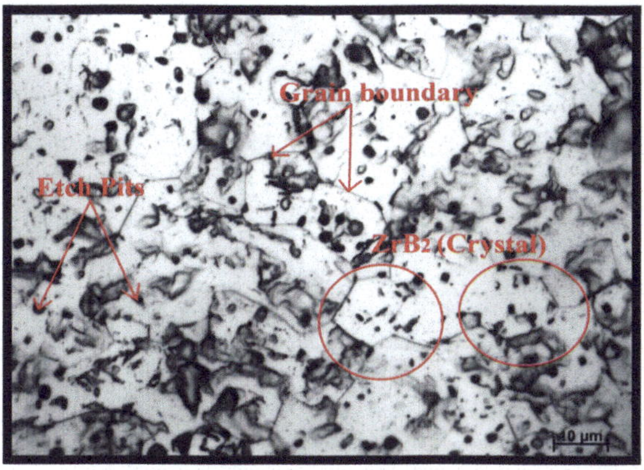

Fig. 2 Optical micrograph of ZrB_2-B_4C(5%) under 100X magnification

The presence of B_4C particles within the crystals may be the cause of uneven crystal size and alignment [10].

The higher magnification image of the SEM micrograph is shown in Fig. 3. The SEM image makes it easy to see the hexagonal ZrB_2 crystal structure, holes that are 1–2 μm in diameter, craters that are 3–4 μm in diameter, and micro-fractures. While machining through WEDM, the material is removed by thermal spalling. At this point, a very high temperature is reached and a sudden cooling occurs, resulting in the accumulation of irregular debris (Fig. 3c) and the formation of a white layer. This debris consists of molten electrode wire, composite oxides, and impurities. As the pulse-on current increases, so does the depth of the heat-affected zone and the thickness of the white layer [11].

Another cause of debris formation is the spalling behavior of the ZrB_2-B_4C during WEDM machining. ZrB_2 can be oxidized during sintering, and the transformation of this ZrO_2 from t-ZrO_2 to m-ZrO_2 while cooling and machining with WEDM causes the re-solidified layer to come off and reveal the spalling surface. When ZrB_2 is oxidized in the atmosphere, ZrO_2 is formed ($2ZrB_2 + 5O_2 = 2ZrO_2 + 2B_2O_3$) [12]. When ZrO_2 gas escapes during the cooling process after sintering, pores of 1–2 μm (Fig. 3a) form on the sample. The hexagonal crystal structure of ZrB_2 can

Table 1 WEDM variable parameters and their levels with min and max ranges

Levels	T_{on}	T_{off}	IP	SV
	(μs)	(μs)	(A)	(V)
1	10	40	120	10
2	14	42	140	20
3	19	45	160	30
4	26	48	180	40
Range of WEDM	1–31	1–63	10–230	1–99
Unit displacement	1	1	10	1

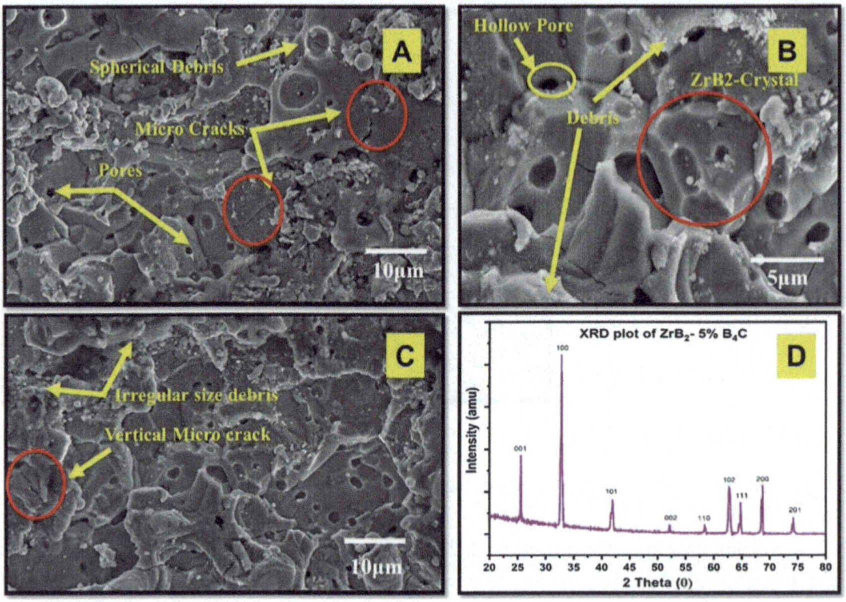

Fig. 3 **a** SEM image of the machined surface of ZrB_2-B_4C (5%) under 2 K magnification; **b** ZrB_2-B_4C (5%) under 5 K magnification; **c** ZrB_2-B_4C (5%) under 2 K magnification; **d** X-ray diffraction plot of ZrB_2-B_4C (5%)

be identified in Fig. 3b. Phase transformation occurs when a zirconium composite is heated to approximately 2100 °C or higher. Tetragonal crystals transform into cubic crystalline structures, which are extremely hard, brittle, and have high compressive strength [13].

Brittleness increases with the number of micro-cracks up until a certain limit is reached. Some cracks of 3–5 μm in diameter and a few craters of 2–3 μm in diameter are observed in the SEM micrography (Fig. 3b). These craters reflected the presence of B_4C material inside them. Cracks may form as a result of thermo-mechanical mismatch and geometric discontinuity, resulting in residual stresses and additional stress concentrations formed during the sintering process and machining.

3.2 X-Ray Diffraction Analysis

For the ZrB_2-B_4C (5 wt.%), the XRD pattern is displayed in Fig. 3d. XRD analysis is done to identify the crystalline phase present in the sample by comparing the observed diffraction peaks with known reference patterns from the International Centre for Diffraction Data (ICDD) or the Crystallography Open Database (COD). The analyzed peaks at (1) 25.56°, (2) 32.64, and (3) 42.76° are high-intensity peaks that indicate the characteristics of crystallographic plane ZrB_2 and B_4C, which often overlapped with ZrB_2. The secondary phases or broad peaks at (1) 52.68° (2) 57.76°, (3) 62.55°, (4) 64.79,°(5) 68.25,° and (6) 74.11° identify the crystal structure and lattice parameters of ZrB_2 and also indicate small crystallite size, defects, overlapping peaks from multiple phases, and impurities. The graph does not explicitly show peaks for B_4C, which might be the existence of low volume fraction of its concentration (less than 5%) or overlapping peaks with ZrB_2. If B_4C peaks are present, they might be less intense and could overlap with ZrB_2 peaks, making them less distinguishable.

3.3 Physical Properties

The bulk density and relative density percentage of hot-pressed ZrB_2–5% B_4C composites SPS at 2200 °C is 5.76299 g/cm^3 and 96.14%, respectively, and Vickers hardness is 1535 HV, which is equivalent to 15.059 GPa. Bulk density of the composite reduces with increasing B_4C percentage [14]. Aside from that, it demonstrates that the hardness of the composite rises with rising amounts of B_4C, which is quite normal considering the incredibly high hardness of boron carbide (30 GPa) [15]. One essential thing to notice here is that the bulk density is significantly low compared to its high hardness, with an exceptionally high melting point. With these characteristics, it is a possible material for space vessels, missiles, and other uses [16].

3.4 Mathematical Modeling and Response Surface Regression Analysis

The goal of machining processes is to create a high-quality, accurate surface finish, and high material removal rate in the shortest amount of time. Using three variable machine parameters, the present research was evaluated with variations of DF, IP, and SV, which are WEDM machining parameters. Tables 1 and 2 show the machining statistics for the current task, as well as the levels that have been established. RSM was used to evaluate machine reaction parameters using Minitab statistical software. To create a mathematical model and forecast the precision of the specified model for response parameters such as machining time, surface roughness, and machining speed. In general, a mathematical model of a second-order polynomial response surface equation is created and explained as follows. Regression equation in un-coded units for machining time, surface roughness, and machining speed.

Machining-time (s).

$$\begin{aligned} &11727 - 97585 \times DF + 111 \times IP + 135 \times SV + 94870 \times DF2 \\ &- 0.393 \times IP2 - 3.42 \times SV2\ 102 \times DF \times IP + 424 \times DF \\ &\times SV\ 1.005 \times IP \times SV \end{aligned} \tag{3}$$

Table 2 WEDM experimental run order table

SL no.	Input parameters			Responses		
	DF	IP(A)	SV (V)	Machining time (sec)	Surface roughness (μm)	Machining speed (mm^2/min)
1	0.20	120	10	84.53	1.68	2.84
2	0.20	140	20	81.14	1.709	2.95
3	0.20	160	30	77.5	1.744	3.096
4	0.20	180	40	54.71	1.984	4.386
5	0.25	120	20	65.4	1.874	3.669
6	0.25	140	10	72.7	1.795	3.301
7	0.25	160	40	49.52	2.012	4.846
8	0.25	180	30	62.62	1.902	3.832
9	0.30	120	30	65.25	1.911	3.678
10	0.30	140	40	53.44	1.982	4.491
11	0.30	160	10	65.6	1.86	3.658
12	0.30	180	20	62.56	1.922	3.836
13	0.35	120	40	42.1	2.104	5.7
14	0.35	140	30	58.52	1.955	4.101
15	0.35	160	20	59.26	1.89	4.049
16	0.35	180	10	60.2	1.89	3.986

Surface-roughness (Ra).

$$2.762 + 10.96 \times DF - 0.01194\, xIP - 0.0087 \times SV - 10.98 \times DF^2 + 0.000049\, IP^2 + 0.000206 \times SV^2 - 0.0170 \times DF \times IP - 0.0327 \times DF \times SV + 0.000084 \times IP \times SV \quad (4)$$

Machining-speed (mm/s).

$$0.179 + 1.667 \times DF - 0.00227 \times IP - 0.00209 \times SV - 1.835 \times DF^2 + 0.000007 \times IP^2 + 0.000055 \times SV^2 - 0.00071 \times DF \times IP - 0.00843 \times DF \times SV + 0.000019 \times IPxSV \quad (5)$$

Speed and surface roughness. According to reports [17], a low T_{off} slows down machining speed as the gap becomes unsteady. The machining efficacy of these ceramic composites with WEDM has been found to be comparable to that of metallic components. However, it only functions with a limited set of machining parameters.

As a result, several tests are conducted to meticulously pick machine parameters for efficient machining of such difficult-to-machine ceramic composites. It has also been noted that the duty factor and servo voltage (SV) have the biggest effects on machining speed, whereas IP has the least impact. As the T_{on} and peak current

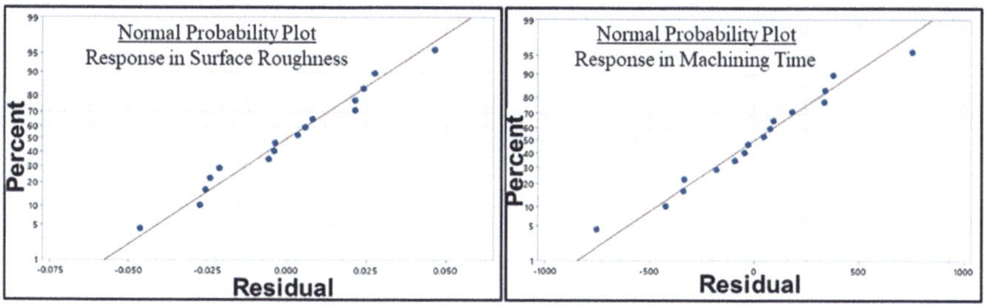

Fig. 4 Normal probability plot illustrating **a** surface roughness **b** machining time

increase, irregular debris and micro-cracks develop, increasing surface roughness [18]. The ordered data is plotted with a rough estimate of the means or medians of the associated order statistics to create the normal probability plot (Fig. 4). It is used to evaluate whether a small collection of data belongs to a normal distribution. In other words, if the diagram yields a straight line, the process is normally distributed [19]. As a result, determining the process capability is a good option.

4 Conclusion

The laminar and homogeneous ZrB_2–B_4C–5%wt composite was fabricated using SPS at 2200 °C under 50 MPa pressure with a 15-minute soaking time. Microstructural analysis and mechanical property evaluation were conducted, followed by machinability testing using WEDM. SEM analysis revealed cracks (3–5 μm) along grain boundaries and craters (2–3 μm) comparable in size to B_4C crystals, likely indicating their presence in cavities. Crack formation along grain boundaries may result from thermomechanical mismatch and geometrical discontinuities due to residual stress from machining and sintering. The XRD plot of the material reveals distinct peaks corresponding to the crystalline phases of ZrB_2 and B_4C present in the material. The sharp and well-defined peaks suggest a high degree of crystallinity in the material. The absence of significant unidentified peaks implies minimal impurity phases, confirming the successful synthesis of the ZrB2–5% B4C composite. The addition of B_4C enhanced ZrB_2's mechanical properties without compromising its intrinsic characteristics, improving hardness while slightly reducing relative density compared to pure ZrB_2. Additionally, the material demonstrated excellent machinability, making it suitable for hypersonic and high-temperature applications. Analytical and experimental findings indicate that as DF increases, machining speed and surface roughness improve, with DF and SV being the primary factors affecting machining speed, while IP has the least influence.

References

1. Kumari A (2015) Impact of humidity and temperature on ceramic materials. Int J Comput Bus Res. http://www.researchmanuscripts.com
2. Monteverde F, Bellosi A, Scatteia L (2008) Processing and properties of ultra-high temperature ceramics for space applications. Mater Sci Eng A 485(1–2):415–421. https://doi.org/10.1016/j.msea.2007.08.054
3. Gupta N, Basu B (2018) Hot pressing and spark plasma sintering techniques of intermetallic matrix composites. In: In intermetallic matrix composites: properties and applications. Elsevier, pp 243–302. https://doi.org/10.1016/B978-0-85709-346-2.00010-8
4. Pramanick A, Mandal S, Pratim Dey P, Kr Das P (2021) WEDM process optimization of sintered structural ceramic sample by using fuzzy-MPCI technique. Mater Today Proc 41:925–934. https://doi.org/10.1016/j.matpr.2020.10.350
5. Saha N, Chakraborty S, Dey PP, Das PK (2014) Machining of ZrB_2-SiC composites by wire-EDM technique. Trans Indian Ceram Soc 73(2):94–97. https://doi.org/10.1080/0371750X.2014.922420
6. Yang F, Zhang X, Han J, Du S (2008) Mechanical properties of short carbon fiber reinforced ZrB_2-SiC ceramic matrix composites. Mater Lett 62(17–18):2925–2927. https://doi.org/10.1016/j.matlet.2008.01.076
7. Zhang X, Xu L, Du S, Han J, Hu P, Han W (2008) Fabrication and mechanical properties of ZrB_2-SiCw ceramic matrix composite. Mater Lett 62(6–7):1058–1060. https://doi.org/10.1016/j.matlet.2007.07.044
8. Chakraborty S, Debnath D, Mallick AR, Das PK (2015) Mechanical, tribological, and thermal properties of hot-pressed ZrB_2-B_4C composite. Int J Appl Ceram Technol 12(3):568–576. https://doi.org/10.1111/ijac.12290
9. Sunilkumar D, Dey HC, Albert SK, Shaikh H, Arvinth Davinci M, Murugesan S (2021) Characterization of interfacial microstructure and mechanical properties of the diffusion bonded CP-Ti to AISI 304L stainless steel. Trans Indian Inst Metals 74(1):9–20. https://doi.org/10.1007/s12666-020-02116-2
10. Chakraborty S, Debnath D, Mallick AR, Gupta RK, Ranjan A, Das PK, Ghosh D (2015) Microscopic, mechanical and thermal properties of spark plasma sintered ZrB_2 based composite containing polycarbosilane derived SiC. Int J Refract
11. Goyal A, UR Rahman H, Ghani SAC (2021) Experimental investigation & optimisation of wire electrical discharge machining process parameters for Ni49Ti51 shape memory alloy. J King Saud Univ-Eng Sci 33(2):129–135. https://doi.org/10.1016/j.jksues.2020.01.003
12. Huang K, Xia Y, Wang A (2023) High temperature oxidation and oxyacetylene ablation properties of ZrB_2-ZrC-SiC ultra-high temperature composite ceramic coatings deposited on C/C composites by laser cladding. Coatings 13(1):173. https://doi.org/10.3390/coatings13010173
13. Prakasam M, Valsan S, Lu Y, Balima F, Lu W, Piticescu R, Largeteau A (2018) Nanostructured pure and doped zirconia: synthesis and sintering for sofc and optical applications. InTech, In Sintering technology–method and application. https://doi.org/10.5772/intechopen.81323
14. Mandal S, Chakraborty S, Dey PP (2022) A study of mechanical properties and WEDM machinability of spark plasma sintered ZrB_2–B_4C ceramic composites. Micron 153. https://doi.org/10.1016/j.micron.2021.103198
15. Pramanick A, Kumar R, Mandal S, Dey PP, Das PK (2018) Modeling and multi-objective optimization of WEDM of spark plasma sintered boron carbide considering preferences of users. In: IOP conference series: materials science and engineering. 377(1). https://doi.org/10.1088/1757-899X/377/1/012094
16. Gopinath NK, Jagadeesh G, Basu B (2019) Shock wave-material interaction in ZrB_2–SiC based ultra high temperature ceramics for hypersonic applications. J Am Ceram Soc 102(11):6925–6938. https://doi.org/10.1111/jace.16548
17. Rajurkar IP, Wang WM (1993) Thermal modeling and on-line monitoring of wire-EDM. In J Mater Process Technol 38

18. Bobbili R, Madhu V, Gogia AK (2015) An experimental investigation of wire electrical discharge machining of hot-pressed boron carbide. Def Technol 11(4):344–349. https://doi.org/10.1016/j.dt.2015.07.001
19. Dada M, Popoola P, Mathe N, Pityana S, Adeosun S (2020) Parametric optimization of laser deposited high entropy alloys using response surface methodology (RSM). https://doi.org/10.1007/s00170-020-05781-1/Published

Response Variation with Electrical Parameters during Slitting of Copper Thin Sheet in Micro-Electrical Discharge Machining

Tanmoy Majumder and Promod Kumar Patowari

Abstract The rising demand for high-precision micro-components in MEMS has led to the development of different micro-fabrication processes. Over the last twenty years, micro-slitting of thin sheets with a particular emphasis on materials used in MEMS has gained a lot of interest. Copper is a highly used material in electronic gadgets, and the slitting operation of copper thin sheets has to be explored thoroughly. In this research work, the micro-slitting and machinability study of copper has been performed by varying the electrical process parameters. Variations in responses such as material removal rate (MRR), overcut (OC), and edge deviation (ED) corresponding to process parameters are also observed. Responses are optimized with the utilization of overall evaluation criteria (OEC), with the best parameters being a voltage of 80 V and a capacitance of 100 pF.

Keywords μEDM · Copper · Micro-slitting · ANOVA · OEC

1 Introduction

The increasing and widespread use of MEMS-based devices has greatly accelerated the trend of miniaturization. The micro-manufacturing business in MEMS is projected to surpass $25 billion during the next ten years [1]. Metals such as Au, Cu, Ag, and Al are used frequently in MEMS-based devices [2]. Cu thin sheets are used to fabricate inexpensive MEMS and diode switches. Copper micro-grooves are often used in motherboards and vacuum electrical equipment [3, 4]. Slit-type openings are widely seen in many MEMS-based devices [5]. MEMS microphones utilize a membrane featuring slit-type structures to provide superior signal-to-noise ratios [6]. Micro-slits in MEMS resistive microphone wafers enable mobility while ensuring pressure drop, maintaining balance, and directing sound waves [7]. Micro-slit machining in metals can be easily achieved through the micro-electrical discharge

T. Majumder (✉) · P. K. Patowari
Department of Mechanical Engineering, NIT Silchar, Silchar, Assam, India
e-mail: tanmoy94majumder@gmail.com

A. Biswas et al. (eds.), *Recent Advancements in Artificial Intelligence and Optimization*, Lecture Notes in Mechanical Engineering,
https://doi.org/10.1007/978-981-95-9268-5_19

machining (μEDM) process. μEDM is a very pertinent and efficient nonconventional micro-machining method owing to its capacity to create complex geometries with precise accuracy at micro-level [8]. Mouralova et al. [9] constructed a precise slit of 5000 × 170 μm in copper, utilizing the same copper foil as a tool. Kliuev et al. [10] demonstrated the potential of micro-shaping nonconductive materials by EDM-milling operations with a μ-electrode having a diameter of Ø = 0.5 mm. This method, however, is characterized by significant burr production and a poor material removal rate (MRR). Moses and Jahan [11] efficiently fabricated micro-features, including micro-holes, micro-slots, and patterns, on titanium grade 5 alloy (Ti-6Al-4 V) with μEDM. They detected debris particles adhering to the surface and margins of the micro-features. D'Urso et al. [12] implemented μEDM drilling on SS and tungsten carbide (WC) utilizing brass and WC electrodes. Their observations indicated that features of tool and workpiece materials, including electrical resistivity, melting point, and thermal conductivity, considerably influence the machining process. They furthermore developed an extensive process index derived from process parameters and material characteristics. Jahan et al. [13] conducted μEDM drilling on WC with a resistance-capacitance (RC) and transistor-type pulse generator. They concluded that the RC-type generator is better-suitable for producing micro-structures with enhanced dimensional precision and surface polish. Sapkal and Jagtap [14] detected an enhancement in MRR during the micro-drilling of Ti-alloys with elevated voltage and capacitance. Singh et al. [15] investigated the optimization of processes and performance attributes in micro-hole drilling. An increase in entrance and exit overcut occurred with the rise in voltage and capacitance. Sahoo et al. [16] explored the influence of process constraints on Ti-diamond by the drilling of micro-holes utilizing μEDM. They noted that the MRR rose with elevated input voltage and capacitance, leading to enhanced discharge intensity and improved material erosion of the workpiece. Kuriakose et al. [17] investigated the influence of machining settings on the accuracy of micro-hole drilling in Zr-based metallic glass. They observed the variation in OC, edge deviation (ED) with voltage and capacitance.

In this research work, micro-slitting has been performed on copper using a WC electrode through μEDM. WC is chosen as the electrode due to its exceptional erosion resistance [18]. The research seeks to study the impact of process variables on responses during the micro-slitting operation.

2 Methodology

A table-top RC-type μEDM setup is used to perform the research. A WC tool of 489.32 μm in diameter is used for the micro-slitting of a copper sheet having a thickness of 100 μm. Figure 1 depicts the μEDM setup and micro-tool used in this research work. Hydrocarbon oil is chosen as dielectric to remove the debris particles produced in the machining process.

The literature [19–22] and pilot experiments suggest that the effect of electrical process parameters (voltage and capacitance) is much higher on responses compared

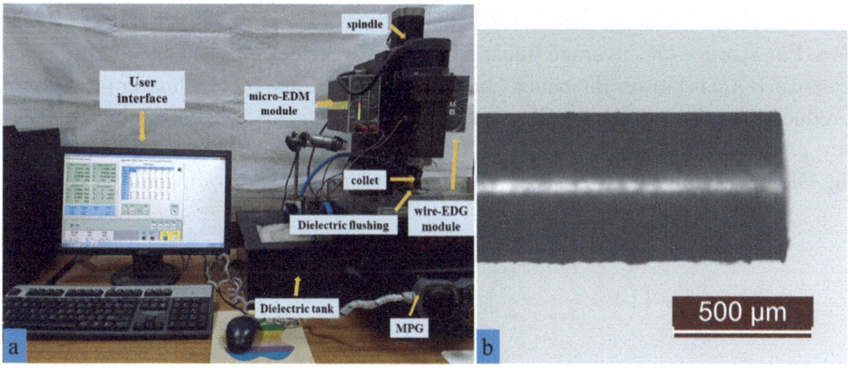

Fig. 1 **a** μEDM setup **b** WC micro-tool

to non-electrical parameters. In this research work, micro-slitting of copper has been performed by varying the electrical parameters and keeping non-electrical parameters constant in the μEDM process. Tool rotation and feed rate are kept constant at 500 rpm and 10 μm/s respectively. Table 1 depicts the process constraints selected for the experimental study.

MRR, average overcut (avg. OC), linear ED, and average circular ED are the responses of this experimental study. MRR is measured in terms of the volume of material removed during the slitting process. Schematics of output constraints of the slitting operation, such as linear and circular ED at start and finish, OC, etc., are shown in Fig. 2.

Table 1 Depiction of process constraints

Parameter	Abbreviation	Quantity	Unit
Capacitance	C	10^2, 10^3, 10^4	pF
Voltage	V	80, 110, 140, 170, 200	Volt

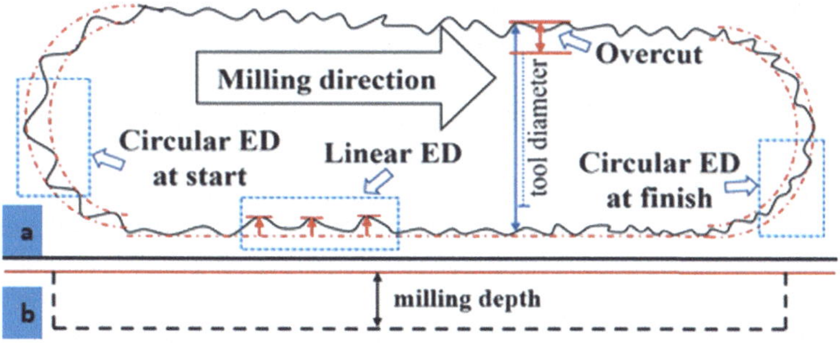

Fig. 2 Schematic appearance of OC, circular ED, and linear ED—**a** top view **b** side view

The overall evaluation criteria (OEC) are selected to optimize the process parameters. For average OC, average linear ED, and average circular ED, the quality characteristic (QC) is "lower the better," however for MRR, the QC is "higher the better." A reduced average OC, linear ED, and circular ED indicate less dimensional inaccuracy, enhancing precision, while a larger MRR signifies an elevated production rate. The OEC is established for each experimental condition and computed using Eq. 1.

$$OEC = \left(\left(1 - \frac{p-p_{\min}}{p_{\max}-p_{\min}}\right) \times W_p\right) + \left(\left(1 - \frac{q-q_{\min}}{q_{\max}-q_{\min}}\right) \times W_q\right) + \left(\left(1 - \frac{r-r_{\min}}{r_{\max}-r_{\min}}\right) \times W_r\right) + \left(\left(\frac{s-s_{\min}}{s_{\max}-s_{\min}}\right) \times W_s\right) \quad (1)$$

where p, q, r, and s represent average OC, average circular ED, average linear ED, and MRR, and Wp, Wq, Wr, and Ws are the weight proportions assigned to p, q, r, and s, respectively.

3 Results and Discussions

μEDM of Cu thin sheet is conducted utilizing a WC micro-tool, with all responses from the full factorial experiment are recorded carefully. Key performance indicators in μEDM that correlate with productivity and process accuracy include MRR, OC, and ED.

Figure 3 depicts the effect of V and C on the response parameters such as MRR, OC, linear, and circular ED. Average circular ED is measured by taking the average of circular ED at start and finish. When the V and C rise, the discharge energy (DE) provided in the inter-electrode gap rises. With the rise in C, the duration of discharge increases in addition to the increase in discharge energy per pulse, causing higher and faster removal of material from the workpiece. Thus, it increases the MRR and OC. Whereas with the surge in DE, the linear and circular ED increase due to the accumulation of higher heat, causing abrupt melting and evaporation of material from the workpiece.

ANOVA is utilized to analyze the process elements that significantly influence performance attributes. According to the results shown in Table 2, capacitance is identified as the most significant process parameter influencing the responses, with voltage following closely after. MRR surges when the V and C surge, which is favorable. However, simultaneously, the surge in ED and OC also requires vigilant assessment. Optimizing process parameters is crucial to establish optimal conditions for all response measures. OEC is designated for optimization in the current investigation. The OEC is calculated by giving equal importance to MRR, average OC, average circular ED, and average linear ED. The highest OEC value denoting the optimum condition is found at a state of voltage = 80 V and capacitance = 100 pF. MRR, average OC, average circular ED, and average linear ED at the optimum

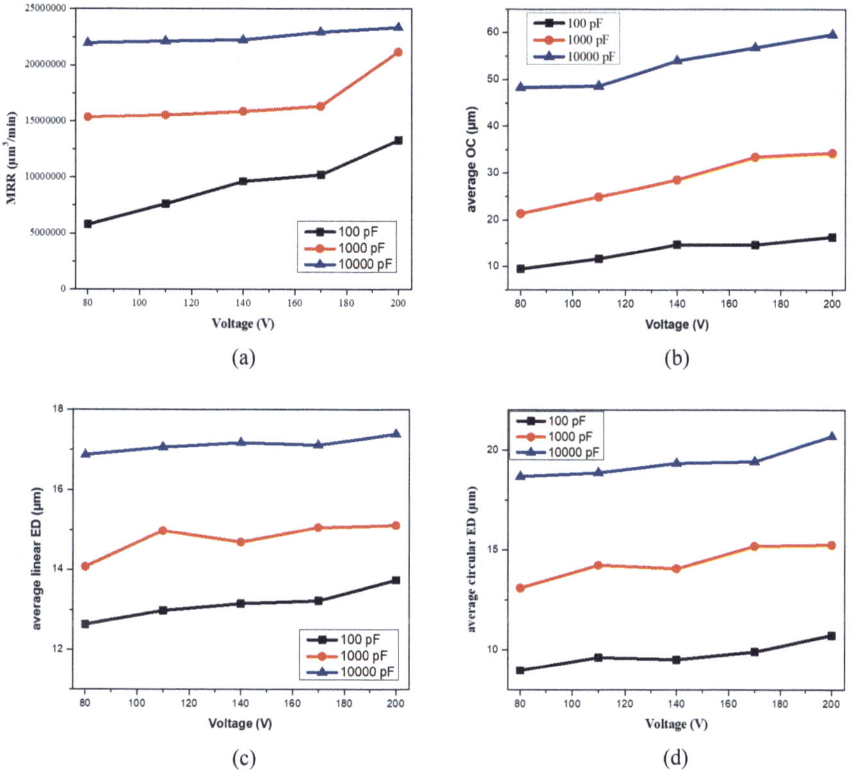

Fig. 3 Variation of responses with input constraints—**a** MRR, **b** average OC, **c** average linear ED, **d** average circular ED

condition are found to be 5795306.81 μm^3/min, 9.46 μm, 8.97 μm, and 12.63 μm respectively. Figure 4 depicts the micro-slit machined at optimum conditions.

4 Conclusions

Based on the obtained findings during the micro-slitting operation, some significant conclusions may be enumerated as follows:

- The quantity of response measures increased with the increase in process variables.
- Although the increase in MRR with the increase in process parameters is desirable but concurrently the increase in other responses hampers the geometrical efficiency.
- The optimum operating condition is found at voltage = 80 V and capacitance = 100 pF.

Table 2 ANOVA of responses

Source	DOF	Seq SS	Contribution	Adj SS	Adj MS	F-value	P-value
MRR							
C	2	4.41×10^{14}	88.52%	4.41×10^{14}	2.2×10^{14}	119.31	0.00
V	4	4.24×10^{13}	8.52%	4.24×10^{13}	1.06×10^{13}	5.74	0.02
Error	8	1.47×10^{13}	2.97%	1.47×10^{13}	1.85×10^{12}		
Total	14	4.98×10^{13}	100.00%				
$R^2 = 97.03\%$, R^2 (adj) $= 94.81\%$							
Average OC							
C	2	4111.14	94.21%	4111.14	2055.57	688.52	0.00
V	4	229.01	5.25%	229.01	57.25	19.18	0.00
Error	8	23.88	0.55%	23.88	2.99		
Total	14	4364.04	100.00%				
$R^2 = 99.54\%$, R^2 (adj) $= 99.09\%$							
Average linear ED							
C	2	40.13	96.34%	40.13	20.06	561.33	0.00
V	4	1.24	2.98%	1.24	0.31	8.67	0.00
Error	8	0.28	0.69%	0.28	0.04		
Total	14	41.66	100.00%				
$R^2 = 99.31\%$, R^2 (adj) $= 98.80\%$							
Average circular ED							
C	2	233.61	96.92%	233.61	116.80	1185.18	0.00
V	4	6.62	2.75%	6.62	1.66	16.80	0.00
Error	8	0.79	0.33%	0.79	0.09		
Total	14	241.02	100.00%				
$R^2 = 99.67\%$, R^2 (adj) $= 99.43\%$							

- Average MRR, OC, circular ED, and linear ED at optimal conditions are 5795306.81 μm^3/min, 9.46 μm, 8.97 μm, and 12.63 μm.

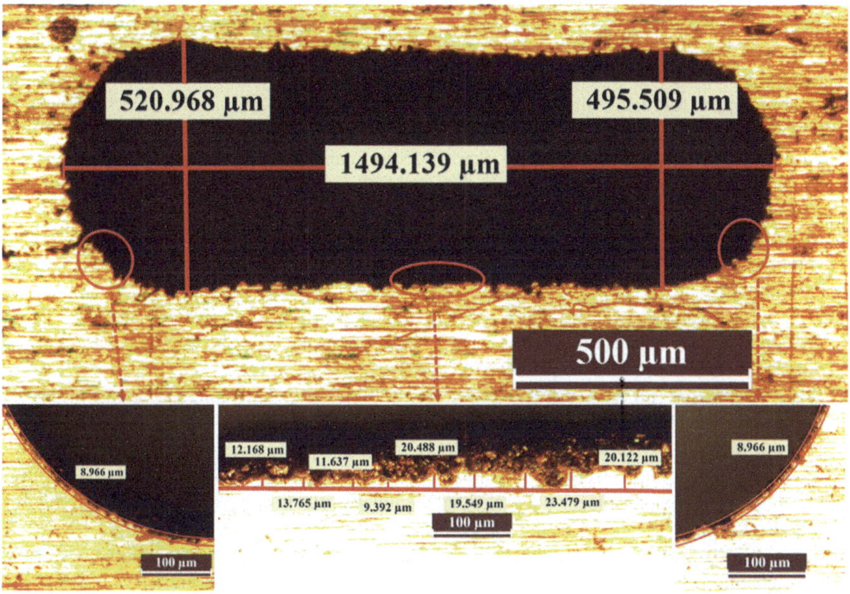

Fig. 4 Microscopic representation of the machined slit at OEC

References

1. Pourabdollahian G, Copani G (2017) Market analysis, technological foresight, and business models for micro-manufacturing. In: Springer tracts in mechanical engineering, pp 261–291. https://doi.org/10.1007/978-3-319-39651-4_10/FIGURES/18
2. Souilah M, Chaabi A, Perdigones F, Quero JM, Flores G, Lain MR (2016) Fabrication process for PCBMEMS capacitive pressure sensors using the cu layer to define the gap. IEEE Sensors J 16:1151–1157. https://doi.org/10.1109/JSEN.2015.2503405
3. Hao X, Xu W, Chen M, Wang C, Han J, Li L et al (2021) Laser hybridizing with micro-milling for fabrication of high aspect ratio micro-groove on oxygen-free copper. Precis Eng 70:15–25. https://doi.org/10.1016/J.PRECISIONENG.2021.01.012
4. Balaraman D, Bhattacharya SK, Ayazi F, Papapolymerou J (2002) Low-cost low actuation voltage copper RF MEMS switches. In: 2002 IEEE MTT-S international microwave symposium digest (cat. No.02CH37278), vol 2. IEEE, pp 1225–1228. https://doi.org/10.1109/MWSYM.2002.1011879
5. Majumder T, Patowari PK (2024) A review on the use of microslit for various applications and its fabrication. J Braz Soc Mech Sci Eng 46:1–23. https://doi.org/10.1007/S40430-023-04638-4/FIGURES/12
6. Yoo I, Sim J, Yang S, Kim H (2018) Development of capacitive MEMS microphone based on slit-edge for high signal-to-noise ratio. In: 2018 IEEE micro electro mechanical system. IEEE, pp 1072–1075. https://doi.org/10.1109/MEMSYS.2018.8346745
7. Nebhen J, Savary E, Rahajandraibe W, Dufaza C, Meillere S, Haddad F et al (2013) Low-noise CMOS analog-to-digital interface for MEMS resistive microphone. In: 2013 IEEE 20th international conference on electronics, circuits, and systems. IEEE, pp 445–448. https://doi.org/10.1109/ICECS.2013.6815450
8. Jahan MP, Wong YS, Rahman M (2010) A comparative experimental investigation of deep-hole micro-EDM drilling capability for cemented carbide (WC-Co) against austenitic stainless

steel (SUS 304). Int J Adv Manuf Technol 46:1145–1160. https://doi.org/10.1007/S00170-009-2167-8/METRICS
9. Mouralova K, Bednar J, Benes L, Plichta T, Prokes T, Fries J (2022) Production of precision slots in copper foil using micro EDM. Sci Rep 12:5023. https://doi.org/10.1038/s41598-022-08957-9
10. Kliuev M, Maradia U, Boccadoro M, Perez R, Stirnimann J, Wegener K (2016) Experimental study of EDM-drilling and shaping of SiSiC and SiC. Procedia CIRP 42:191–196. https://doi.org/10.1016/j.procir.2016.02.269
11. Moses MD, Jahan MP (2015) Micro-EDM machinability of difficult-to-cut Ti-6Al-4V against soft brass. Int J Adv Manuf Technol 81:1345–1361. https://doi.org/10.1007/S00170-015-7306-9/METRICS
12. D'Urso G, Maccarini G, Ravasio C (2016) Influence of electrode material in micro-EDM drilling of stainless steel and tungsten carbide. Int J Adv Manuf Technol 85:2013–2025. https://doi.org/10.1007/S00170-015-7010-9/METRICS
13. Jahan MP, Wong YS, Rahman M (2009) A study on the quality micro-hole machining of tungsten carbide by micro-EDM process using transistor and RC-type pulse generator. J Mater Process Technol 209:1706–1716. https://doi.org/10.1016/j.jmatprotec.2008.04.029
14. Sapkal SU, Jagtap PS (2018) Optimization of micro edm drilling process parameters for titanium alloy by rotating electrode. Procedia Manuf 20:119–126. https://doi.org/10.1016/j.promfg.2018.02.017
15. Singh AK, Patowari PK, Chandrasekaran M (2020) Experimental study on drilling micro-hole through micro-EDM and optimization of multiple performance characteristics. J Braz Soc Mech Sci Eng 42:1–14. https://doi.org/10.1007/S40430-020-02595-W/FIGURES/10
16. Sahoo R, Kar S, Patowari PK (2020) Experimental study on titanium diamond by fabricating micro-holes using micro-EDM with Taguchi's method and response surface method. Eng Res Express 2:45026. https://doi.org/10.1088/2631-8695/ABCC46
17. Kuriakose S, Patowari PK, Bhatt J (2019) Effect of micro-EDM machining parameters on the accuracy of micro hole drilling in Zr-based metallic glass. Eng Res Express 2:15001. https://doi.org/10.1088/2631-8695/ab5c72
18. Liu W, Jia Z, Zou S, Zheng X (2014) A new measurement method of relative volume wear ratio based on discharge debris composition analysis in micro-EDM. https://doi.org/10.1155/2014/479609
19. Majumder T, Kebede AW, Patowari PK, Sahoo CK (2024) Micro electrical discharge drilling in SS304 and its multiparameter optimization using desirability function analysis. Eng Res Express 6:25569. https://doi.org/10.1088/2631-8695/ad56fd
20. Majumder T, Patowari PK (2025) A study on machinability of Ti-6Al-4V ELI (grade 23) titanium alloy using micro-EDM process. Springer, Singapore, pp 297–303. https://doi.org/10.1007/978-981-97-4167-0_25
21. Kar S, Sarmah P, Baroi BK, Patowari PK (2020) Drilling of micro-holes in titanium using micro-EDM: a parametric investigation. In: Lecture notes in mechanical engineering, pp 589–600. https://doi.org/10.1007/978-981-15-0124-1_53/FIGURES/10
22. Alavi F, Jahan MP (2017) Optimization of process parameters in micro-EDM of Ti-6Al-4V based on full factorial design. Int J Adv Manuf Technol 92:167–187. https://doi.org/10.1007/S00170-017-0103-X/METRICS

The manufacturer's authorised representative in the EU is Springer Nature Customer Service Centre GmbH, Europaplatz 3, 69115 Heidelberg, Germany. If you have any concerns regarding our products, please contact ProductSafety@springernature.com

Printed and bound by CPI Group (UK) Ltd, Croydon, CR0 4YY

21/09/2026

02219991-0005